THE POLITICS OF CULTURAL POLICY IN FRANCE

The Politics of Cultural Policy in France

Kim Eling
Associate Expert
International Labour Office
Geneva
Switzerland

First published in Great Britain 1999 by
MACMILLAN PRESS LTD
Houndmills, Basingstoke, Hampshire RG21 6XS and London
Companies and representatives throughout the world

A catalogue record for this book is available from the British Library.

ISBN 0–333–74669–4

First published in the United States of America 1999 by
ST. MARTIN'S PRESS, INC.,
Scholarly and Reference Division,
175 Fifth Avenue, New York, N.Y. 10010

ISBN 0–312–21974–1

Library of Congress Cataloging-in-Publication Data
Eling, Kim, 1970–
The politics of cultural policy in France / Kim Eling.
p. cm.
Includes bibliographical references and index.
ISBN 0–312–21974–1 (cloth)
1. France—Cultural policy. 2. France—Politics and government—1981–1995. 3. Art and state—France. 4. Art patronage–France. 5. Mitterrand, François, 1916– —Influence. I. Title.
DC423.E44 1999
306'.0944—dc21 98–43275
CIP

This book is printed on paper suitable for recycling and made from fully managed and sustained forest sources.

10 9 8 7 6 5 4 3 2 1
08 07 06 05 04 03 02 01 00 99

Printed and bound in Great Britain by
Antony Rowe Ltd, Chippenham, Wiltshire

To Gerburg and Ludger Eling

Contents

List of Tables and Figures

Tables

Figures

Acknowledgements

It would be impossible to thank all of the many individuals on both sides of the Channel whose help made this book possible. Above all, I owe an inestimable debt to my PhD supervisor at the London School of Economics, Howard Machin, without whose support I could never have embarked on the thesis on which *The Politics of Cultural Policy in France* is based. The form and substance of the book owe much to his subtle but never less than inspiring comments and suggestions – and to his lucid but unfailingly sympathetic approach to deciphering the theatrical world of French public life.

My thanks are also due to Professor Peter Hall of Harvard University, and to Alain Guyomarch at LSE, for their comments and encouragement. Of the members of staff in the European Institute and Government Department at LSE who also offered help and advice at different stages of the project, I particularly wish to thank Jean-Charles Lagrée for letting me partake of his extensive network of contacts in Paris, and Mark Thatcher for advice on the theoretical aspects of my work. I also wish to thank the staff and students at various LSE research seminars for their valuable comments on draft versions of individual chapters.

I further wish to extend my thanks to the many officials and cultural professionals in France who made this project possible by agreeing to recount their experiences of the *années Lang*. While it would not be possible to list all those individuals who, by their *disponibilité* and generosity, made my fieldwork a tremendously rewarding and stimulating experience, my particular gratitude must go not least to Jack Lang himself, but also to Michel Chauvière, Jean-Jacques Hocquard, Jean-Michel Lemoyne de Forges, Pierre Mayol, Pierre Moulinier and Marc Nicolas for their help in providing essential *introductions* to the arcane worlds of the French administration and arts establishment. Access to past and present *hauts fonctionnaires* of the Ministry of Culture was greatly facilitated by the assistance of the *Association des anciens élèves de l'Ecole Nationale d'Administration*.

Much of the initial research for this project was carried out in the library of the *Département des Etudes et de la Prospective* of the Ministry of Culture, to whose staff I extend my gratitude.

Finally, my thanks are due to the Economic and Social Research Council for funding my research over three years, and notably for its generous support of a year's fieldwork in Paris.

List of Abbreviations and Acronyms

AFPF	Association Française des Producteurs de Films
AJT	Action pour le Jeune Théâtre
APUR	Atelier Parisien d'Urbanisme
ARP	Société des Auteurs, Réalisateurs, Producteurs
CDN	centre dramatique national
CIR	Centre d'Information du Rock
CNC	Centre national de la cinématographie
CNSMP	Conservatoire National Supérieur de Musique et de Danse de Paris
COIC	Comité d'organisation de l'industrie cinématographique
CRC	centre régional de la chanson
DAGEC	direction de l'Administration générale et de l'Environnement culturel
DDC	direction du Développement culturel
DDF	délégation au Développement et aux Formations
DMD	direction de la Musique et de la Danse
DRAC	direction régionale des Affaires culturelles
DTS	direction du Théâtre et des Spectacles
EAC	établissement d'action culturelle
EIC	Ensemble InterContemporain
EPPV	Etablissement Public du Parc de la Villette
FAIR	Fonds d'Action et d'Initiative Rock
FIC	Fonds d'intervention culturelle
FRAC	Fonds régional d'art contemporain
FRAM	Fonds régional d'acquisitions des musées
IFCIC	Institut pour le Financement du Cinéma et des Industries Culturelles
IRCAM	Institut de Recherche et de Coordination Acoustique/Musique
PS	Parti socialiste
PSU	Parti socialiste unifié
SNAC	Secrétariat national à l'action culturelle
SOFICA	Société pour le Financement de l'Industrie Cinématographique et Audiovisuelle

SYNDEAC	Syndicat national des directeurs d'entreprises artistiques et culturelles
TSA	Taxe Spéciale Additionnelle
UPF	Union des Producteurs de Films

Introduction

The national policy process in the French Fifth Republic is widely assumed to exemplify a statist pattern of decision-making: policy is seen as being initiated and defined by the state, and extra-governmental groups are perceived as playing a passive or subordinate role. However, our knowledge of the reality of relations between the French state and 'societal' interests remains strikingly patchy. The prevalence of the statist paradigm in official rhetoric and the legalistic tradition in French political science have militated against empirical analysis of the roles played by public officials and private interest groups in the elaboration and implementation of specific policies.

The enterprise of rolling back the 'monolithic' view of the French state is, of course, well under way. An extensive and iconoclastic literature has been devoted to the relationship between local political, administrative and economic elites since the 1960s.[1] Moreover, detailed empirical studies of the relationship between the administration and its 'clients' have been undertaken in a small number of 'technical' policy sectors, such as Keeler's analysis (1987) of agricultural policy-making, or Suleiman's account (1987) of the influence of the French notarial profession. Studies of industrial policy-making such as those by Bauer and Cohen (1981), Cohen (1989), Feigenbaum (1985), Hayward (1986b) or Padioleau (1981) have similarly suggested that the ability of policy-makers to develop and impose a coherent set of policies may in many sectors be subverted by the influence of economic and professional interest groups both within and outside the state. However, the nature of policy-making in many of the most symbolically important areas of state intervention has barely been subjected to systematic analysis.

Not surprisingly, the policy areas most strongly associated with the practice of Fifth Republic presidentialism – of which foreign and defence policy are the more frequently cited examples – have remained most consistently insulated against empirical scrutiny. The policy area considered in this book impinges less obviously on the wellbeing and survival of the French nation. Nevertheless, cultural policy is widely seen to have moved into a presidential *domaine réservé* since 1958: perhaps in view of the status accorded culture and cultural debates in French society, and against the background of a long tradition of direct state intervention in the cultural domain, successive Presidents of the

Fifth Republic have not only taken an interest in, but sought directly to influence the cultural life of the French nation.

The tendency to consider cultural policy an object of direct political decision-making, and this at the highest levels of government, is not, however, an innovation of the Fifth Republic. In France, the significance of cultural policy as an attribute of national government has traditionally exceeded the narrow confines of what, in Anglo-Saxon countries, is more modestly termed 'arts policy'. In its aims, if not in the actual content of policy, French cultural policy has long been invested with a higher 'national' mission in addition to the prosaic function of supporting individual artists and cultural institutions. In consonance with an historical 'republican' consensus about the need to defend the integrity of the French nation, complemented in a more contemporary period by concern about the integrity of the French language, policy towards the arts has always been conceived of as part of a wider 'nation-building' enterprise.[2] At a more narrowly political level, cultural policy has periodically assumed a 'regime-building' role in periods of constitutional or ideological change. Thus, decisive phases in the institutional development of a national cultural policy have often coincided with periods following major regime changes; the creation of the Ministry of Cultural Affairs in 1959, following the establishment of the Fifth Republic in 1958, is the most obvious case in point.[3]

Perhaps as a function of the association of cultural policy with higher political or national imperatives, the allocation of arts funding even at the level of individual institutions has traditionally been subject, in France, to hierarchical political decision-making by elected politicians or their representatives – in stark contrast with the Anglo-Saxon convention of delegating funding decisions to 'intermediary' institutions like the British Arts Council. Moreover, major decisions bearing on the organization of France's cultural institutions have been considered the preserve of the highest levels of government, a convention accentuated since the inception of the Fifth Republic.

No President – and, indeed, few rulers of pre-Republican France – went further than François Mitterrand in asserting this prerogative. Under no other President, moreover, did the Minister of Culture and his administration occupy as important a place, both in terms of the prestige and public recognition they enjoyed, and in terms of the financial means at their disposal. What is more, the Socialist years were marked, perhaps more than any other period of contemporary French history, by strong partisan and ideological influences – or, at the very least, aspirations – in policy-making. Crucially, these coincided with the presence of an exceptionally homogeneous parliamentary majority and of a President whose hegemony was undisputed within the governing party – in short, a conjunction of factors which might be expected to consolidate and amplify the strength and unity of the French state *vis-à-vis* 'societal' actors. There can be no better period, then, to explore the practice of cultural policy-making in contemporary France, and the extent to which a statist, and more specifically presidentialist, picture of the policy process can be upheld, than the decade, from 1981 to 1986 and from 1988 to 1993, during which a Socialist President ruled alongside a Socialist government.[4]

The question that this book seeks to answer by subjecting cultural policy-making in the Mitterrand years to a detailed analysis is a simple one: Who 'governs' the arts in France? In other words, is the theoretical and rhetorical omnipotence of the state a reliable indicator of the actual processes by which policy priorities are set and by which funds are allocated? And to what extent are actors outside the institutional boundaries of the state able to shape the policies which affect them? The only way of sketching answers to these questions consists in attempting to reconstitute, chronologically, the preparation, elaboration and implementation of individual decisions, and in charting the role played by different actors at each step of the process. To do so even for the most momentous policy choices made across the entire range of the activities of the Ministry of Culture would be impossible. The approach chosen in this book is more modest in its scope. It focuses on a small number of sectors and policy initiatives, in which the policy process is considered, primarily on the basis of interviews with key policy-makers – a term taken here, crucially, to encompass not only politicians, their advisers and the bureaucracy, but also a wide variety of extra-governmental actors. The findings from the case studies are taken – boldly, but pragmatically – as representative of a more generally prevailing cultural 'policy style'.

After a brief historical overview of the evolution of arts funding policy in France, in Chapter 1, and a presentation, in Chapter 2, of the analytical tools and theoretical frameworks which may help answer the question 'Who governs?' in more rigorous terms, the empirical evidence is tackled in the six chapters that make up the core of this book. Chapter 3 tells the story of the *Cité de la Musique*, one of the architectural *grands projects* that came to further beautify the French capital during the 'reign' of François Mitterrand – the story told is not, however, the 'presidential' tale that the monarchic airs surrounding the *grands projects* might have led one to expect. Chapter 4 explores policy-making in the contemporary music sector, an area of the performing arts that may seem far removed from the stately realm of the *grands travaux* – but which in fact involves a similar set of actors, and presents roughly the same distribution of roles, as the previous chapter. Chapters 5 and 6 deal with policy-making towards, respectively, the theatre and film production, two large and long-established sectors in which state intervention plays a crucial role – and in which, as the narrative of both chapters suggests, the beneficiaries of public funding have become highly adept at making their voice heard in government. Chapters 7 and 8, by contrast, cover two areas that were more marginal to the cultural policy of the Socialist governments – the Lang Ministry's unorthodox foray into the popular music sector, and policy on *développement culturel*, or cultural development. A concluding chapter seeks to collate the findings thrown up by the sectoral case studies; to develop, if not a model, at the very least a wider picture of the kind of factors that determine the influence of different actors in different areas of cultural policy; and in so doing point up some of the limitations of readings of the French polity which rely too heavily either on the study of constitutional articles or on a regalistic rhetoric which the representatives of the state are themselves only too happy to indulge in. The last aspiration is, of course, the most ambitious: however, this book is written firmly in the belief that only by chiselling away at small corners of the edifice that is the apparent monolith of the French state can we hope to do justice to the richness and complexity of French political life.

1 Culture and the State in France

In few liberal democracies does government policy towards the arts occupy as prominent a position in political discourse, or does government involve itself as directly and as intimately in the cultural life of a nation, as in contemporary France. Indeed, many countries have chosen to devolve responsibility for arts funding to quasi-autonomous 'arm's length' institutions: not so France, where both Presidents and Ministers have amply used their powers to create cultural institutions, to allocate funding among existing ones, and to bestow favours on individual artists. The institutional arrangements of the Fifth Republic have heightened the tendency for cultural affairs to be perceived as a matter for the highest echelons of political power. In a regime often remarked upon for its near-monarchic traits, the power to dispense not only political, but also artistic patronage, and to leave a lasting mark on the cultural life of one's 'subjects', has come to be seen as a natural prerogative of the 'republican monarch'. Yet the conventions of the Fifth Republic only reinvigorated and perpetuated a tradition firmly anchored in French political history: if interpreted broadly, 'cultural policy' could be said to have been invented by France's absolutist monarchy, and subsequent regimes only built on the absolutist core, adding layer upon layer of later priorities and preoccupations – traces of which persist to this day.

1.1 STATE INTERVENTION IN CULTURE SINCE THE *ANCIEN RÉGIME*: BUILDING A TRADITION OF PATRONAGE AND *DIRIGISME*

The monarchic tradition in French cultural policy can be summarized as centred on two complementary roles for the state: first, that of patron, offering direct support to individual artists and institutions; and second, that of censor, concomitantly imposing tight controls on the production and distribution of works of art. The foundations for both roles were laid in the seventeenth century. In 1635, under Louis XIII, Richelieu thus instituted the *Académie française* as guardian of an 'official' conception of art; in 1667, the Academicians held their first official exhibition – a precursor to the *Salon* which would act as showpiece of an 'official' art from 1725 onwards. At the same time, the monarchy developed its role as patron of the arts. In 1680, under Louis XIV, the *Comédie française* was thus created as an 'official' theatre company – a status it has in some ways preserved to this day, since its *Administrateur* is appointed by decree in the Council of Ministers. In the decorative arts, too, the state began to acquire a dominant, and in some areas exclusive, role as patron, *via* the creation of the royal *manufactures*. The reign of Louis XIV also saw the appointment of what may be termed France's first 'Minister of Culture', with the designation of Colbert as 'Surintendant des Bâtiments, Arts et Manufactures'.

The institutions created by the absolutist regime were shaken only at the margins by the Revolution. While restrictions on artistic freedom were briefly eased, the *Directoire* and the Napoleonic empire quickly reimposed both censorship and the pre-eminence of the official academies: throughout the nineteenth century, public support for the arts would thus be marked by the official recognition accorded to a clearly circumscribed group of 'academic' artists, and by the material encouragement granted these official artists by means of public commissions or work on projects such as the *Opéra Garnier*. Moreover, the government's direct role in the running of individual cultural institutions would be continually reinforced. As the scope of public intervention expanded, so did the number of institutions effectively controlled by the state *via* its powers not only to determine their budgets, but also to appoint and dismiss their artistic directors – a system that has survived to the present day in public cultural institutions such as the five *théâtres nationaux*.

The first significant deviation from the tradition of the state as patron and censor coincided with the creation of the Third Republic in the 1870s. While the monarchic and academic strands in cultural policy had emphasized only the goal of artistic 'excellence', however defined, as intrinsically valuable, the cultural ambitions of successive governments of the Third Republic introduced another policy objective into their arts funding policy – a goal concerned not with the act of artistic creation, but with the widening of access to culture, or the 'democratization' of culture. While a small number of initiatives introduced in the years after the Revolution – such as the opening of the Louvre as a museum in 1793, or the plan, mooted in 1791, to establish a public museum in each *département* – had already set precedents for a policy of cultural 'democratization', the reign of an academic conception of art that persisted until the end of the Second Empire had prevented their further development. Under the Third Republic, by contrast, arts policy was linked explicitly with the policy of 'popular education': support for the arts now came to encompass the encouragement of artistic education, both in the *école républicaine* which emerged as the Third Republic's greatest achievement, and in specialized municipal music or art schools. For the first time, moreover, the aim of bringing professional artistic 'excellence' to a larger audience was stated – and, on a very modest scale, realized with the creation of institutions like the *Théâtre national populaire* in 1920. Institutionally, too, the years of the Third Republic contributed greatly to the emergence of a cultural policy conceived as such – in 1870 and again in 1881, a Ministry devoted specifically to the arts was created, though on both occasions its competences quickly passed to the Ministry of Education, where they remained for most of the Third Republic period.

The Third Republic's concern with 'popular education' culminated in the efforts of the left-wing Popular Front government of 1936–8; even the Popular Front failed, however, fundamentally to change the *de facto* priorities of cultural policy – which continued to accord pride of place to 'excellence', even if the goal of 'democratization' was more and more openly proclaimed. Subsequently, under the Vichy regime of 1940–4, the authoritarian aspects of the monarchic tradition in French cultural policy once again came to the fore – although it should be noted that not all of Vichy's actions in the cultural domain were disavowed after the *Libération*: thus, the institution set up by Vichy to control and 'corporatize' the French film industry, the *Comité d'organisation de l'industrie cinématographique* (COIC), would serve after the war as a model for the agency responsible for the administration of film policy, the *Centre national de la cinématographie*.

The first years of the Fourth Republic saw a renewed emphasis on the goal of cultural 'democratization' – at least at the level of aspirations and declarations. The Preamble to the 1946 Constitution thus declared: 'The Nation guarantees equal access for children and adults to schooling, to professional training and to culture'.[1] The concrete achievements of the Fourth Republic were less exalted. Nevertheless, two innovations of the post-war years deserve mention, to the extent that they would condition the policies of subsequent governments: first, the creation, from 1944 onwards, of local *maisons des jeunes et de la culture*, precursors to Malraux's *maisons de la culture*; and the policy known as 'décentralisation théâtrale', or theatrical decentralization –

which aimed to establish an extensive and accessible network of public theatres throughout France. The administrative organization of public support for the arts remained too fragmented, however, for a self-conscious 'cultural policy' to emerge: throughout the Fourth Republic, arts funding was the responsibility of a *direction générale des Arts et Lettres* attached to the Ministry of Education, and while the Fourth Republic's ephemeral governments occasionally included a junior Minister for the *Beaux Arts*, none of these, and none of the Fourth Republic's Presidents, left a lasting imprint on the French arts funding system. The creation of an explicit and ambitious 'cultural policy' was to be left to the Ministers and Presidents of the Fifth Republic.

1.2 CULTURAL POLICY IN THE FIFTH REPUBLIC: MINISTERS AND POLICIES

The evolution of national cultural policy from the foundation of the Ministry of Cultural Affairs in 1959 to the Socialists' accession to power in 1981 can be divided into three basic periods.[2] The first, 'founding' period of 1959 to 1969 under André Malraux, appointed the first Minister of Cultural Affairs shortly after the inception of the Fifth Republic, was marked by an emphasis on the institutional dissemination of 'high' culture, perceived as a means of furthering the Gaullist project of national integration. In a second, brief period from 1969 to 1974, shaped by the tenure as Minister of Jacques Duhamel, policy was characterized by a greater emphasis on a less institutional, 'socio-cultural' form of cultural 'democratization'; in contrast with the Gaullist period, efforts to widen access to the arts were focused on small-scale, community-based projects. A third phase, congruent with the Giscard presidency, saw a decline in the ambition of 'democratization', conceived either in the paternalistic 'Malrassian' mould or in a 'socio-cultural' sense; concurrently, there was a substantial decline in the resources available to the Ministry.

Cultural Policy under Malraux: 'Democratization' and its Limitations

The policy pursued under Malraux can be interpreted as an attempt to combine the goal of cultural 'democratization' with a traditional concept of artistic 'excellence': that is, cultural policy was perceived, at least at the level of idealized objectives, as a process of bringing 'high' culture to 'the people'. Thus, the Ministry of Cultural Affairs was committed by its founding decree to 'making the great masterworks of mankind, and above all of France, accessible to the greatest possible number of Frenchmen, providing the largest possible audience for our cultural heritage, and encouraging the creation of works of art and of the spirit which will enrich it' (quoted in Wangermée, 1988, p. 29). Malraux's chosen instrument for this task was a network of multidisciplinary institutions, the *maisons de la culture*, one of which was to be built in every large town in France. However, both the feasibility of the kind of cultural 'democratization' envisaged by Malraux, and the adequacy of the resources devoted to its advancement, were questionable.[3] The priority accorded to the goal of 'democratization', or *action culturelle*, in the Minister's rhetoric was only partially matched by substantive policy initiatives. Only twelve *maisons de la culture* were ever built, and the budgetary credits allocated to Malraux's Ministry amounted, on average, to no more than 0.39 per cent of the overall government budget. Public funding continued, moreover, to be directed overwhelmingly at bastions of 'high' culture; thus, in 1968, almost a decade after the foundation of the Ministry of Cultural Affairs, the subsidies to two established Parisian institutions, the Paris Opera and *Opéra-Comique*, accounted for one quarter of the Ministry's overall budget.[4] Indeed, Malraux's attempt to reconcile the opposing poles of 'democratization' and 'excellence' in the *maisons de la culture* was effectively abandoned in the 1970s and 1980s. The *maisons* having failed, on the whole, to widen arts audiences, many of them mutated into conventional arts venues in which the professional imperative of artistic 'excellence' prevailed.

Cultural Policy in the 1970s

The policies of Malraux's successors were, on the whole, no less ambiguous, and failed equally to resolve the contradiction between 'democratization' and 'excellence'. Jacques Duhamel, Minister of Cultural Affairs from 1971 to 1973, has generally been credited with making the most determined effort to expand the Ministry's remit beyond the institutional sphere of 'high' culture, notably through the creation of a new policy mechanism, the inter-ministerial *Fonds d'intervention culturelle* (FIC), which was to provide funding for 'innovative' small-scale projects not covered by the Ministry's established sectoral funding mechanisms or by the rigidly institutional form of *action culturelle* established by Malraux. However, Duhamel's initiatives made a lasting impact only at the margins of the Ministry's policies. Subsequently, particularly during the presidency of Valéry Giscard d'Estaing, both the policy of *action culturelle* and the political and budgetary status of the Ministry were to regress significantly. Thus, the Ministry of Cultural Affairs was formally demoted to the status of mere *secrétariat d'Etat* on Giscard's accession to the presidency, and the share of cultural expenditure in the state budget, which had risen to 0.61 per cent in the last year of the Duhamel Ministry, gradually declined to 0.47 per cent by 1981. Throughout the Giscard years, the Ministry's overall priorities remained vague: Giscard's four *secrétaires d'Etat* were either indifferent or hostile to the concept of *action culturelle* as defined, and redefined, by Malraux and Duhamel, and none of them had a coherent alternative set of policy objectives; consequently, the Giscard years were widely seen as a period of stagnation in French national cultural policy.[5]

Cultural Policy after 1981 (I): Goals and Ideology

On coming to power in 1981, the French Socialists had no shortage of elaborate designs for the future shape of cultural policy. At the level of general political objectives, the 1972 *programme commun* of the Socialist and Communist Parties contained an extensive section on cultural policy, comprising both a critique of the previous governments' record and prescriptions for a future left-wing government, as did the 1980 *projet socialiste* and François Mitterrand's ultimately more influential *110 propositions* of 1981. Taken together, these documents pointed both to a more complete ideological analysis of the relationship between culture, government and the economy and to a more coherent and comprehensive package of specific policies than any previous government had offered or set itself the task of implementing. The *programme commun* had defined the objective of a cultural policy of the Left as 'breaking the stranglehold of the financial powers on all cultural activities: film, theatre, books, records, television, video tapes, etc.' (quoted in Gournay, 1988, p. 310). François Mitterrand had, as presidential candidate in 1981, described his policy priorities in similar terms: 'No to a policy of profitability, to audience figures and to opinion polls, to the concentration of decision-making powers; yes to the allocation of real funding' (quoted in Gournay, 1988, p. 309). In other words, the Socialist Party (PS) situated itself very clearly as tending towards a view on which 'culture', at least in a narrow sense, and the 'economy', in the widest possible sense, were quite explicitly to be divorced. Another important feature of the Socialists' pre-1981 declarations on cultural policy was the emphasis on *création*, or professional cultural production. This may seem obvious, not least since the conservative governments of the 1970s were accused by the PS of having starved France's cultural institutions of resources. It is nonetheless significant in that it could be seen to commit the PS to a policy goal potentially inconsistent with the ideal of cultural 'democratization', however defined. The *110 propositions* included the goal of reinforcing *animation*, or measures aimed at audience participation in cultural activities, and *diffusion*, the wider distribution of cultural goods.[6] Yet they did so without establishing a clear order of priorities, thus leaving open the prospect of a fundamental break with the policies of previous Ministers, and a reinvigoration of *action culturelle* – without, however, committing resources to any particular policy option.[7]

This ambiguity became more pronounced in the public presentation of the Socialists' cultural policies after the appointment of Jack Lang as Minister of Culture in 1981. The overall ambitions of the Lang Ministry are – somewhat paradoxically, given the radical content of the Socialists' *projet* for French society in the early 1980s – difficult to specify; indeed, the absence of a single, overarching policy goal may be an important feature of the 'Lang years', as may be the absence of clear precepts as to how relations between government and *les créateurs* were to be organized.[8] Moreover, although his public declarations seemed to prioritize the vague ideal of *création*, Lang rejected the notion that the Ministry's policy had to express a choice between the conflicting goals of artistic 'excellence' and widening access, and suggested the two should be put on an equal footing:

> I would be tempted to reply that I do not want to choose; I do not want to be trapped in this infernal dialectic.... It is absolutely necessary to avoid coming down on one side or the other of the question. It is necessary at the same time to encourage high-quality artistic education and to facilitate artistic activity in all its forms, across all social classes. (quoted in Clément, 1983, p. 299)

Nevertheless, there were generalizable 'themes' in the Lang Ministry's public presentation of its policy goals, which present an interesting contrast with the theses put forward by the Socialists before 1981. Among the most important of these 'themes' was an implicit rejection of the universalist aspirations that had characterized Malraux's conception of the Ministry's goals, with its attendant distinction between 'high' and 'low' culture. Indeed, many of the Ministry's most prominent initiatives during the 'Lang years' could be seen as attempts to expand the definition of culture as such to include virtually any activity expressive of a particular 'lifestyle'. Related to this, the Ministry's official discourse implied a rejection of the belief – central to Malraux's conception of the aims of cultural policy – that the state should be concerned with making a certain kind of culture more widely accessible, since it denied the value judgments about different forms of expression implicit both in Malraux's paternalistic vision and in later forms of *action culturelle*.[9]

Malraux and his successors had openly expressed their disdain even for traditional popular art forms like *chanson* – indeed, Malraux had in one notorious incident ejected the owner of the Olympia concert hall in Paris from his office after the latter had had the temerity to request some form of public recognition for the art of *chanson*. Lang, by contrast, not only accepted popular culture as in no way inferior to 'high' culture, but actively sought to expand the remit of his Ministry to encompass the entirety of the artistic forms, 'high' or 'low', that happened to be practised in France in the early 1980s. In a series of symbolic and highly public gestures, Lang thus extended 'official' recognition – in the form of financial support – not only to traditional French forms of popular culture such as *chanson* or the circus arts, but also, and more controversially, to forms of expression originally imported from Anglo-Saxon popular culture – such as rock music and strip cartoons, or *bande dessinée*. In concrete terms, this involved, for example, allocating small start-up grants to young French rock groups, or supporting the creation of a *bande dessinée* 'museum' in the town of Angoulême – though the funding involved never amounted to more than a fraction of the subsidies allocated to institutional 'high' culture.

As a corollary to the rejection of a rigid distinction between 'high' and 'low' culture, and perhaps surprisingly for a Socialist Minister of Culture, Lang also effected both a 'reconciliation' between the public funding system and the large sphere of cultural production not dependent on public subsidy, and a recognition of the commercial dimension of the cultural sector as a whole. At a practical level, this meant a reinforcement of the various direct and indirect funding mechanisms aimed at the 'cultural industries' – that is, the sectors of cultural production operating largely or entirely on a commercial, profit-making basis.[10] Again, this represented a significant departure from the premises of cultural policy in the Malrassian sense, in which the delivery of culture to the people was conceived of fundamentally as a public service, and forms of cultural production surviving only through the interplay of supply and demand in the market economy were perceived almost by definition as mere 'entertainment'. Indeed, some observers have argued that Lang effectively brought about 'the dissolution of cultural policy *via* its legitimation on economic grounds' (Urfalino, 1993b, p. 848).[11] Yet as with the distinction between 'high' and 'low' culture, the limits of Lang's innovative approach become apparent when the Ministry's policy is considered not only in the light of public declarations and symbolic gestures, but as reflected in actual funding commitments and budgetary priorities – where 'serious' funding remained the preserve, in most sectors, of institutions purveying 'high' culture and insulated against the laws of supply and demand.

Cultural Policy after 1981 (II): Funding Patterns

In absolute terms, the Socialists' accession to power made an immediate and dramatic impact on central government spending on culture. The budget of the Ministry of Culture for 1982 represented an increase

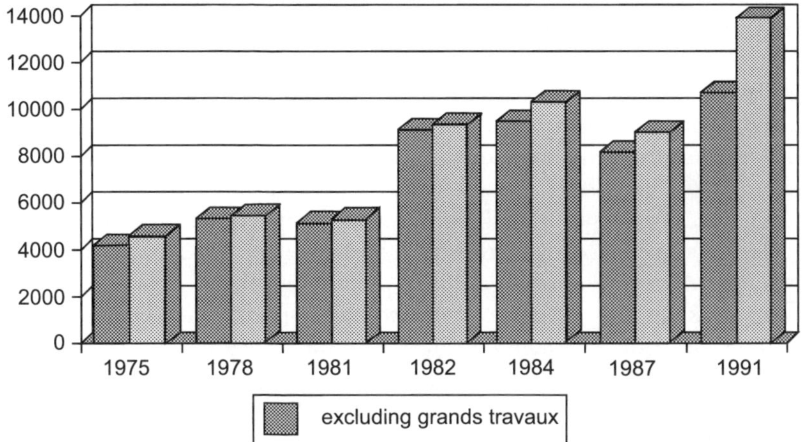

Figure 1.1 Ministry of Culture: *dépenses ordinaires* (current expenditure) and *autorisations de programme* (projected capital expenditure), 1975–91 (millions of francs indexed to 1991)

Source: *Ministère de la Culture.*

in real terms of 79 per cent over the previous year's budget,[12] corresponding to a rise in the share of cultural expenditure in total central government expenditure from 0.47 per cent to 0.76 per cent.[13] The upward trend was maintained in later years, though much of the subsequent increases was absorbed by the capital projects known as the *grands travaux* (see Figure 1.1).

The implications of the way the new funds were distributed among different sectors of cultural production, meanwhile, were more ambiguous. *Développement culturel*, a broad category covering a wide variety of measures aimed specifically at cultural 'democratization' and artistic education, benefited from the largest rise in budgetary allocations, with an increase in real terms of 1,440 per cent. It should be added, however, that the 1981 budget of *développement culturel* had amounted to only 41 million francs, and it is worth noting that the budget for *développement culturel* regressed continuously in subsequent years (see Figure 1.2). By way of comparison, the budgets for major sectors like drama or music remained largely constant after the near-doubling of 1981. The previously 'neglected' visual arts received an increase of 123 per cent, accompanied by the creation of a separate *délégation aux arts plastiques* in the Ministry's administrative structure; support for film and audiovisual production was increased by 220 per cent.[14] Yet even 'traditional' sectors like drama and music benefited from substantial increases in their budgets, of 75 per cent and 50 per cent respectively – and thus continued to receive the lion's share of public funding for the arts.[15] There was little change in the distribution of funding between different sectors in subsequent years: significantly, only the sector of *développement culturel* was subject to significant budgetary fluctuation after 1982 – an issue we will consider in greater detail in Chapter 8.

The evolution of the Ministry's budget after 1982 confirmed the traditional sectorization of cultural policy: significant policy initiatives were subsequently introduced within each sector, yet little effort was made to introduce a degree of intersectoral co-ordination. As Menger (1990, p. 114) argues, types of intervention which had previously been seen as 'competing' and inconsistent could now 'coexist' on a more equal basis. This fact should not, however, obscure the equally significant budgetary consolidation, in absolute terms, of traditionally dominant sectors like music and drama. Saez (1988, p. 107) suggests in this context that 'the elitism–populism debate has been resolved only very provisionally by Jack Lang's cultural policy. Lang organized, on a symbolic level, the recognition of cultural diversity and of its social roots, while at the same time reserving his funding for "professional

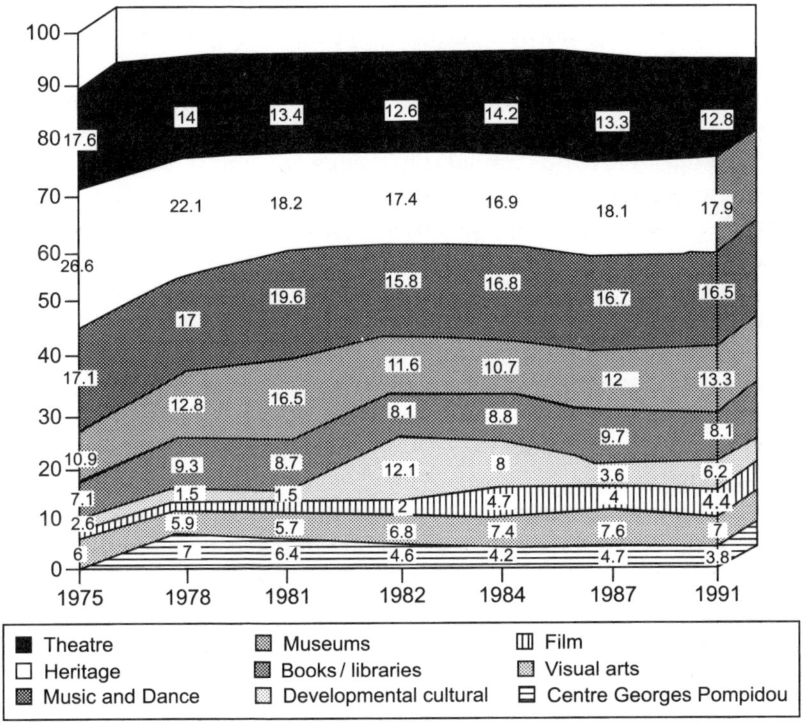

Figure 1.2 Ministry of Culture: breakdown of annual budget, by sector, 1975–91 (*Dépenses ordinaires* and *autorisations de programme*, % of overall spending excluding *grands travaux*).

Source: *Ministère de la Culture*.

quality"'. In other words, the budgetary priorities established by centuries of *Beaux-Arts* policy remained unshaken. New and 'popular' art forms were admitted into the institutional fold: yet they remained marginal adjuncts to an edifice constructed over time by France's official and implicit artistic *académies*.

The grands travaux: cultural policy as a 'politics of grandeur'?

The policies pursued by the Socialists in government embraced the entirety of the cultural policy traditions that successive French regimes and governments had secreted since the eighteenth century. Alongside the concern with *Beaux Arts* 'excellence' and the goal of 'democratization', a further strand was readily assimilated into the Socialists' cultural policy: that is, the construction of prestigious public edifices at the initiative or under the patronage of the Head of State, known as the *grands projets* or *grands travaux*. Rooted in the absolutist tradition by which French monarchs sought to immortalize themselves by building palaces and gardens or, as in the case of Napoleon III, redesigning the layout of an entire capital, the *grands travaux* were 'rediscovered' from the 1960s and 1970s onwards by the Presidents of the Fifth Republic. Georges Pompidou thus inaugurated a long line of recent presidential projects with the arts centre that was to bear his name; Giscard is associated with projects such as the science museum at La Villette or the Musée d'Orsay. However, no President used his powers to initiate and direct large-scale architectural projects – the *fait du prince* – as extensively as François Mitterrand, and buildings such as the Louvre pyramid and the *Grande Arche de la Défense* stand as perhaps the most durable testimony to Mitterrand's two seven-year presidential terms.

Of course, the *grands travaux* were not exclusively 'cultural' in their inspiration or function. Only some of the projects initiated during the Mitterrand presidency were to be put to explicitly cultural purposes: the Louvre pyramid, the Opéra Bastille, the *Cité de la Musique* and the *Très Grande Bibliothèque* most obviously fall into this category; while the *Grande Arche de la Défense* or the new building of the Ministry of Finance at Bercy obviously do not. Moreover, to the extent that they served at least in part as memorials to an individual, the treatment as objects of 'cultural policy' in a narrow sense even of the 'cultural' *grands projets* could be questioned.

However, the conventional link between the *grands travaux* and the 'republican monarch' who stands at the apex of the Fifth Republic's institutional pyramid should not obscure the lack of any direct constitutional or legal foundation for the characterization of the *grands travaux* as essentially 'presidential'. Regardless of the reality of the policy-making process that may have prevailed, the 'cultural' *grands projets* were constructed under the administrative supervision of, and on the budget of, the Ministry of Culture – though with the participation of co-ordinating organs like the inter-ministerial *mission*, and later the *secrétariat d'Etat*, for the *grands travaux*, which were partly responsible for steering the construction programme throughout the Mitterrand presidencies. Furthermore, the completion of projects such as the *Cité de la Musique* and the Opéra Bastille, apart from imposing a virtually perpetual surcharge on the running budget of the Ministry of Culture, had important implications for the Ministry's revenue-funding activities in the performing-arts sectors directly concerned. The stance adopted in this book thus consists in taking at least those *grands projets* with explicitly cultural aspirations and functions as an integral part of the cultural policy of the French Socialists in power. Indeed, the question whether the *grands travaux* are to be perceived as essentially 'presidential' in character, and thus to be treated as standing apart from the more banal everyday politics of the Ministry of Culture, will be taken as entirely empirical – and the conclusions thrown up by the analysis of policy-making with regard to one of the *grands projets*, the *Cité de la Musique*, presented in Chapter 3, are by no means such as to confirm the conventional wisdom.

1.3 CULTURAL POLICY AND LOCAL GOVERNMENT: THE PERSISTENCE OF THE CENTRALIST TRADITION

The *grands travaux* programme could be seen as only the most ostentatious instance of the continuity with prior traditions of French cultural policy, dating back to the *Ancien Régime*, which the Socialists perpetuated. Yet vestiges of the absolutist conception of the role of the state in the cultural domain, which had in a sense been rehabilitated by the creation of a Ministry of Culture in 1959, were also evident in the Socialists' approach to relations with actors outside the central state – and notably the various levels of local government, the *collectivités locales* or *collectivités territoriales*. Indeed, the policy of the Socialist governments towards local government in the cultural policy arena stood in marked contrast to the programme of institutional decentralization that was described, at the time, as *la grande affaire* of François Mitterrand's presidency. While the reforms initiated by Gaston Defferre in 1982 brought about a fundamental change in the relationship between central and sub-national levels of government, and entailed a vast transfer of competences from Paris to the regions, *départements* and communes, the dominant role of the Ministry of Culture in France's cultural life emerged not only undiminished, but even reinforced from the decentralization process. The decentralization laws of the early 1980s scarcely touched upon cultural policy, with no clear delineation of the responsibilities of different layers of government; the only exceptions were the transfer of responsibility for the departmental public libraries, the *bibliothèques centrales de prêt*, to the *départements*, and the attribution of specific competences to the communes with regard to the declaration of conservation areas in the decentralization laws of January and July 1983.[16] It is worth noting that Jack Lang, as Minister of Culture, was vehemently opposed to any form of institutional decentralization.[17] In this, Lang was supported by members of the parliamentary *majorité*, who considered decentralization particularly 'sensitive' in the area of culture, not least because of the Right's electoral ascendancy in the communes, which became more marked with the 1983 municipal elections.[18] Lang's stance also found a favourable echo among the Ministry's artistic constituency: at least some of the Ministry's regional clients were concerned that their autonomy and long-term funding security would be jeopardized by a devolution of funding responsibilities to local politicians. Similar concerns prevailed among senior civil servants at the Ministry of Culture.[19]

The absence of any significant move towards institutional decentralization in the cultural policy area contrasted not only with the policy of the Socialist government as a whole. It also went against the grain of a long-term shift in funding from central government to the *collectivités locales* which had set in as early as the 1970s – long before the Socialists came to power. Indeed, in purely quantitative terms, any analysis of public policy towards the arts in France since the 1960s which fails to accord pride of place to the expansion of cultural spending by the local, departmental and regional authorities might be seen as ignoring one of the most striking developments in French public policy in this domain since the creation of Malraux's Ministry of Cultural Affairs. The most dramatic expansion occurred in the cultural budgets of the communes, by far the most active, in the area of arts funding, of the three types of *collectivité locale*: between 1978 and 1987, spending on culture by France's communal authorities almost doubled, from 8.8 billion francs to 16.3 billion francs – and thus came to surpass even the volume of funding allocated to culture by central government itself.

However, the dramatic development of sub-national funding policies in the 1970s and 1980s should not be seen as implying a 'retreat' of national policy-makers. While the budgetary neglect of national cultural spending under Giscard did underline the increasing weight of sub-national policies, the near-doubling of the resources of the Ministry of Culture in the 1982 budget reversed the dominant financial trend of the previous decades. In 1981, central government funding amounted to 38.7 per cent of total public spending on culture, with the communes accounting for 52.2 per cent of total expenditure; by 1984, the share of central government had risen to 46 per cent, while the communes' share had declined to 47 per cent.[20] More important, the

Ministry succeeded in preserving, and indeed enhancing, its role as the pivot of the funding system, capable both of influencing patterns of cultural production in the sectors it supported, and of shaping the priorities of other funding bodies, notably at the sub-national level.

Thus, while the funding balance between the Ministry and the *collectivités locales* had shifted unambiguously in favour of the latter, the Ministry was still able in many cases to influence the financial decisions of local policy-makers by tying them into institutional mechanisms devised and dominated by itself. The most important of the new funding devices introduced at a sub-national level in the 1980s – notably, two types of visual arts acquisition funds, the *fonds régionaux d'art contemporain* (FRAC) and the *fonds régionaux d'acquisitions des musées* (FRAM) – were devised at a senior level in the Ministry of Culture and effectively grafted onto the policies of the *collectivités territoriales* – which were expected to co-fund the new institutions, with the Ministry retaining a decisive say in artistic decisions through its representatives in the field.[21] Moreover, the Lang years saw a substantial strengthening of the role of the Ministry's regional directorates, the *directions régionales des Affaires culturelles* (DRAC). The creation, in 1977, of the *directions régionales des Affaires culturelles* was intended initially to provide the Ministry with a network of administrative conduits in the regions and facilitate contact with the *collectivités territoriales*. Increasingly, the DRAC emerged as actors in their own right, with the formal authority to dispose autonomously of 'territorialized' funds, or *crédits déconcentrés*, allocated by the Ministry. After 1981, the resources and personnel available to the DRAC were increased substantially, and in 1986, the *crédits déconcentrés* amounted to approximately 9 per cent of the Ministry's total expenditure.[22] Moreover, the DRAC became important interlocutors for the *collectivités territoriales*, in many cases providing technical support services for local policy-makers – and though they came to be seen less and less as mere enforcers of a Parisian writ, the DRAC ultimately reinforced the dependency of local policy-makers on the guidance of the central state.[23]

The indirect influence of the Ministry was evident, too, in the complex financial arrangements to which many 'provincial' institutions and projects were subject. In 1988, it was estimated that two-thirds of all funds dispensed by the *collectivités territoriales* were devoted to projects co-financed by different levels of government, either in the form of uncoordinated 'overlapping funding' (*financements croisés*) or in the form of deliberately concerted 'joint funding' (*financements conjoints*).[24] In the case of the regions, the proportion of funds spent on jointly financed projects rose to an estimated 90 per cent. While there was in most cases no obvious rationale underlying the way in which funding responsibilities were split between different partners, the participation of the Ministry of Culture was widely perceived as a vital 'quality guarantee' for the project or institution concerned – not least by local policy-makers themselves. This enabled the Ministry in many cases to exercise substantial leverage while making only a modest financial commitment: as a former head of the Ministry's theatre directorate observed, 'even small central government subsidies have an extraordinary multiplying effect in many regions. People sometimes complain that there are grants of only 40,000 francs, when in fact these are often tripled or quadrupled in the regions' (Abirached, 1988, pp. 115–16).

The determination of the Ministry of Culture to retain its influence *vis-à-vis* the *collectivités locales* was similarly evident in its 'contractual' policy, which pre-dated the Lang Ministry, and which was a variant of a mechanism commonly used in other areas of central–local intergovernmental relations.[25] In 1974, the Ministry began signing *chartes culturelles* with individual local partners, which, against specific financial and qualitative undertakings on the part of local policy-makers, transferred some of the Ministry's funds to the *collectivités locales*. Owing to budgetary constraints, only 27 *chartes*, with very limited financial implications, were signed, and the scheme was effectively abandoned in 1979.[26] However, a more ambitious contractual policy was initiated by the Lang Ministry in 1981. In part, this took the form of inserting cultural clauses into the general regional development agreements, the *contrats de plan Etat-régions*, concluded in the context of the IXth Plan;[27] simultaneously, the Ministry introduced more specific *conventions de développement culturel,* to be signed with willing *collectivités territoriales*, and again setting out targets and objectives in return for cofinancing by the Ministry.

The use of policy mechanisms such as the *conventions* and the reinforcement of the DRAC should not, of course, be interpreted as a simple recentralization or reversal of previous tentative moves towards devolution. The attitude of sub-national policy-makers was, as noted earlier, anything but uniformly hostile, not least since the political and administrative elites in the *collectivités locales* frequently lacked – by their own admission – both the experience and the political will to undertake 'expansive' policies, in the sense of experimenting with new modes of support or financing activities outside their traditional purview. Crucially, the Ministry of Culture thus retained a dominant position within the funding system, drawing on a range of new funding mechanisms, many of which were developed during the tenure of Jack Lang, and benefiting from the financial leeway it had gained with the budgetary expansion of 1982. In other words, the most important elements of the centralist French cultural policy tradition were only marginally affected by the process of decentralization – underlining, again, both the exceptional symbolic status of cultural policy in the French polity and the extent to which the policies of the French Socialists in power were marked by the influence of concepts and attitudes inherited from earlier periods in the evolution of French cultural policy.

CONCLUSION

Relations between the state and the arts world in France are marked – in appearance, at least – by a tradition of centralized hierarchical policy-making, the origins of which can be traced back to the absolutist monarchs of the seventeenth and eighteenth centuries. The advent of the modern republican state did not alter the basic elements of the 'monarchic' tradition. Rather, successive regimes and governments from 1871 onwards took on its centralistic and occasionally authoritarian traits – and gradually imparted new priorities, notably a concern with 'democratization', to the old system. The Socialist governments of 1981–6 and 1988–93 were no exception: both the *grands travaux* of François Mitterrand and the virtual exclusion of cultural policy from the Socialists' decentralization programme stand as testimony to the strength of the element of continuity in the policy of the Lang Ministry.

However, there are strong *prima facie* reasons for doubting whether the 'monarchic', statist picture of cultural policy-making was ever entirely appropriate. Even without considering the detailed mechanics of policy-making, a brief survey of the evolution of cultural policy outputs, such as that presented in this chapter, may suggest grounds for scepticism. Perhaps we can take the 'monarchic' character of initiatives like the *grands projets*, in which the hand of particular rulers was particularly visible, at face value – though as the findings of the case study of a *grand projet* discussed in this book will suggest, even this may be taking one step too far. In any case, there is no reason to suppose that the same policy style should equally apply to the more 'prosaic' policy sectors which in fact accounted for a much larger share of public spending on the arts.

More important, one of the most striking, and paradoxical, conclusions to emerge from an historical overview of cultural policy in France is a discrepancy, particularly evident since the creation of a Ministry of Culture in 1959, between Ministers' stated policy priorities and the actual distribution of funds between different sectors and projects. Even in the Ministry's founding period, under Malraux, there was a marked disjunction between ministerial rhetoric and budgetary reality: much though Malraux professed his belief that it was the foremost task of a Ministry of Culture to make the great masterworks of humanity accessible to the entire French nation, specific efforts to this end, such as the establishment of the *maisons de la culture*, ultimately remained marginal when set against the *Beaux Arts* 'core' of cultural policy which Malraux inherited from his predecessors. The contrast was no less pronounced during Jack Lang's tenure as Minister. Where Malraux had called for cultural 'democratization', Lang, going further still, called for an extension of the very notion of 'culture', which was now to comprise forms of expression which Malraux could never have conceived of as in any way related to his pantheon of 'great masterworks of mankind'. Yet the cultural policy of France in 1993 much resembled that which the Socialists found when they took over the reins of power in 1981: small amounts of funding had been made available for 'new' and 'popular' artforms, but they remained at best symbolic, at worst risible when set against spending on institutional 'high' culture – as, more importantly, did expenditure on artistic education and measures aimed specifically at widening arts audiences.

The upshot is that, paradoxically, for all its regal trappings and for all the rhetorical attachment to an activist policy of cultural 'democratization', the French arts funding system does not diverge dramatically from what we might have expected to find if France, like the Anglo-Saxon countries, had devolved responsibility for the allocation of public spending on the arts to an 'arm's length' body removed from direct political influences. Of course, it remains to be established whether the hierarchical prerogatives of political decision-makers left a more definite mark on policy in individual sectors. Only by looking at the way in which decisions were taken in individual sectors can we hope to answer this question. More important, only a detailed analysis of sectoral policy-making processes, by identifying the constraints on hierarchical policy-making, and by throwing light on the role of actors beyond the narrow circle of the Presidency and Ministry of Culture, can help us account for the gap between rhetoric and reality in French cultural policy. To facilitate this task, and in order that we may conduct our analysis within a rigorous comparative framework, we now turn to the conceptual tools developed to this end by a wider political science literature.

2 Cultural Policy and Political Science

The study of cultural policy has not been among the traditional priorities of political scientists – a fact which should not surprise us, given the relative financial insignificance of arts funding when set against the more 'vital' responsibilities of government. The development of public support for culture has not generated the theoretical superstructure that has been engendered, say, by the expansion of post-war European welfare states. In consequence, the literature on which we might draw in approaching the subject of cultural policy-making, and against which we might subsequently compare our findings, is extremely limited. Indeed, in order to establish a more coherent framework of analysis for our empirical case studies, we will need to draw on a wider range of theoretical approaches. After considering existing accounts of cultural policy-making in France, this chapter will thus set out and assess a number of more general theories of the policy process in liberal democracies, with particular reference to the study of French public policy-making. In so doing, we will be aiming, in particular, to address the limitations of the different accounts of cultural policy-making previously discussed. It will be our objective not so much to introduce 'models' which can then be tested, using our case studies, but, rather, to develop analytical 'tools' which we can apply to the empirical cases under consideration. Throughout the core empirical chapters of this book, but particularly in deriving conclusions from the case studies, the concepts introduced in this chapter will then be used both as hermeneutic tools to isolate the underlying dynamics of a series of complex processes, and to identify the implications of our findings for the wider debates to which our analysis can be seen as contributing.

2.1 CULTURAL POLICY: EMPIRICAL ACCOUNTS AND THEORETICAL PERSPECTIVES

Cultural policy and policy-making have received scant attention from political scientists. However, the form and functions of cultural policy have traditionally been more widely and publicly debated in France than in any other western democracy, perhaps as a function of the exceptional status accorded cultural policy within the French polity. The years following the accession of the French Socialists to power in 1981 saw a particular surge in public interest in the cultural attributes of the French state: the near-doubling of cultural spending by central government in 1981–2, and the very visible initiatives launched by Jack Lang and François Mitterrand in the field of culture, thus served to rekindle interest in an area of public policy which had been prone to perceptions of 'decline' since the departure of André Malraux from the Ministry of Cultural Affairs. Consequently, the 1980s saw the publication of a wide range of books and articles relating both to cultural policy in general, and to the specific policies of the Lang Ministry.

Much of the material published falls into one of two categories which are of at best limited relevance to an analysis of the policy-making process. The first category comprises polemical essays, many of which were directed as much against the Minister of Culture personally as against the policies of the Ministry.[1] The second category consists of journalistic accounts, which do not, in general, offer either the depth of analysis or the comparative scope to serve as substantive sources with regard to the policy process.[2] Nevertheless, the perceived 'renaissance' of an 'activist' cultural policy also stimulated an unprecedented degree of academic interest in relations between government and the arts, and at least some of the work carried out by French social scientists since the early 1980s merits consideration. The literature on cultural policy and policy-making can be subsumed under three broad headings: first, general descriptions of the cultural policy arena, which tend to follow closely the formal institutional structures of the French funding system; second, accounts of policy-making in specific sectors of public intervention, which tend, however, to focus on very technical issues to the detriment of wider political explanations; and third, 'systemic' accounts, which take a 'global' view of the evolution of cultural policy outputs, but which similarly tend to disregard the explicitly political dimension of cultural policy-making.

Cultural Policy-making in France: General Accounts

Few existing studies of French cultural policy are devoted exclusively to the policy-making process: the elaboration and setting of policy is usually amalgamated with general discussions of policy outputs, and the resulting observations tend to be impressionistic and inadequately documented. However, to the extent that it is possible to discern a clear pattern of arguments about the policy process, two distinct strands of analysis can be identified. The first 'theme' in the literature relates to a distinction between 'high' and 'low' spheres of cultural decision-making. The cultural policy arena is seen, in other words, as divided, with the most prestigious institutions falling within a presidential *domaine réservé*, and less prestigious institutions and sectors coming under the jurisdiction of the Minister of Culture. Thus, Harrison (1993, p. 190), contrasting the *grands travaux* with 'conventional' areas of cultural policy, notes that

> de Gaulle's successors have chosen to mark out a domaine réservé of major cultural projects where their writ has run supreme; other aspects of cultural policy have remained largely under orthodox ministerial control. Lines of responsibility have remained reasonably clear and patterns of decision-making relatively settled.

The second strand in the literature points to the role played by the cultural bureaucracy in shaping the policy priorities of the Ministry of Culture. Thus, in an analysis of the internal administrative structures of the Ministry after 1981, Saez (1988) underlines the tension between a political impetus to introduce new 'forms of management', to 'modernize' the Ministry's central administration under Mitterrand and Lang, and the resistance to such reforms on the part of senior officials. In particular, Saez suggests, the rigidity of existing administrative structures affected the implementation of the Socialists' decentralization strategy and, closely linked with this, the policy of *développement culturel*. On the one hand, Saez argues, the initial flexibility and 'missionary strategy' of the cultural development directorate, the *direction du Développement culturel* (DDC), created by the Socialists in 1982, quickly gave way to a more conventional and institutionalized administrative style; on the other, the devolution of funding responsibilities simply transposed the 'customary rigidity in relations between the world of culture and the administration' to a lower territorial level (Saez, 1988, p. 97). A similar analysis is put forward, with regard to the DDC, by Colin (1990, p. 346), who notes that

> [the DDC] had claimed to be able to shift the policies of the other directorates in order to make them more responsive to the demands of the new, democratically-elected government. This was unacceptable – and all the more so since the working methods of the DDC constituted a break with administrative traditions, particularly through its innovative activities 'in the field'.

Sectoral Accounts of Cultural Policy Formation

The role of civil servants is similarly underlined by more detailed accounts of policy-making and implementation in individual funding sectors. Moreover, while such research remains limited in its scope and presents only a very fragmented picture of the overall policy-making process, sectoral accounts have also had the merit of bringing into focus the central importance in the policy process of relations between officials and their professional 'clients'. Thus, in an account of the evolution of the Ministry's policy in the public libraries sector between 1981 and 1986, Lassalle (1993) emphasizes the role of informal contacts between professional groups, civil servants and elected politicans both in shaping the declared policy aims of the Socialist Party prior to 1981 and in conditioning the choice of policies implemented after the Socialists' accession to power. Conversely, Lassalle points to the limiting effects of professional involvement in policy-making on the degree of policy innovation – thus, while contacts between policy-makers and affected interests facilitated the implementation of expansionary policies in the libraries sector after 1981, they also made it difficult for policy-makers to implement policy changes which would have challenged the prerogatives of the professions affected.

Four (1993), applying a similar analysis to the visual arts sector – specifically, the *Fonds régionaux d'art contemporain* (FRAC) established in the first years of Jack Lang's tenure as Minister of Culture – also points to the close relationship between public and private actors as central to the evolution of a sectoral policy. Noting that the FRAC, which were initially established with a 'democratizing' objective, increasingly adopted the criteria and values prevalent in the professional art world, Four concludes that

> it is no longer possible to distinguish the different actors in the visual arts world according to their public or private-sector background, because the professional practices of the different actors evolve without any real link to their original status. Indeed, it is apparent that each institution or actor 'positions himself' by accepting or by rejecting the values and the professional techniques of the arts world. (Four, 1993, p. 113)

Lassalle's and Four's case studies both point to the central role of professional interests in shaping sectoral cultural policies – either, as in the case of policy towards public libraries, at an early stage in the formulation of policy, or, as in the case of the FRAC, in the implementation of policy. However, the limitations of an analysis based exclusively on a single sector are apparent: for want of a wider research base, we are unable to derive a pattern of professional influence in the policy process applicable to the area of cultural policy as a whole, or to identify the factors which determine the nature and extent of professional influence in the policy-making process.

'Systemic' Accounts of Cultural Policy Formation

Sectoral analyses of cultural policy-making are handicapped by their narrow focus. The third type of approach which can be discerned in the literature on cultural policy, which we might term 'systemic', suffers from no such limitation. The label 'systemic' is taken in this context to designate a variety of approaches, mostly associated not with political science but with organizational sociology, which focus on the study of the cultural sector as an autonomous 'system' of organizations. Though disparate in their precise empirical objects, such 'systemic' approaches, exemplified by DiMaggio's work on the American arts funding system (DiMaggio, 1978, 1986), Friedberg and Urfalino's research into the funding policies of the French *collectivités locales* (Friedberg and Urfalino, 1983, 1984a, 1984b, 1985), Menger's studies of the French contemporary music *milieu* (Menger, 1983, 1989a) and Urfalino's more theoretical and comparative work (Urfalino, 1989), all point to one central feature of arts funding policies in liberal democracies. Thus, the form and content of government funding for the arts was found, in all of the cases studied, to be largely impervious to explicitly political or partisan influences. The institutional development of arts funding is subject, in other words, to what Urfalino (1989) terms 'dynamiques endogènes'. Drawing on Friedberg and Urfalino's findings with regard to French local government arts funding, Urfalino argues that the distributive decisions that constitute a cultural policy have increasingly become detached from political decision-making criteria. Given that the performing arts are subject to an endemic process of cost inflation,[3] Urfalino argues, 'the only room to manoeuvre that remains for political and administrative institutions lies in budget increases' (Urfalino, 1989, p. 92).

The 'endogenous' policy patterns identified by the 'systemic' approach to the analysis of cultural policy provide a promising point of departure for our enquiry. Urfalino's 'dynamiques endogènes' appear to find confirmation in the overview of French cultural policy in the Fifth Republic which we presented earlier, and further underline the tension between policy change and continuity as an issue which any account of the policy process in this area will need to address. However, the 'systemic', sociological approach also leaves a number of crucial questions unanswered. Perhaps most important, it does not explain why the expansionary logic described by Urfalino is sustained on the part of those actors in whom control over budgets and subsidies is ultimately vested – in other words, it does not account for the role, in the policy process, of elected politicians and civil servants, or engage with the nature of the relationship between public officials and their professional 'clients'. Nor does the 'systemic' approach suggest ways of explaining why different institutions, groups of professionals or sectors may be accorded very different levels of public support in a given funding system. In short, we need to explain both why politicians and administrators support particular institutions, and why some professional 'clients' are more successful than others in obtaining public support.

In order to develop a framework which may serve in answering these questions, and which will avoid some of the failings of the very partial accounts discussed above, we will consider a small number of general approaches to the study of the policy process in liberal democracies, drawn essentially from an 'Anglo-Saxon' literature on public policy-making. We will relate these as far as possible to specifically French debates, and to some of the issues which we have identified as specific to the area of cultural policy.

At a very general level, two contrasting strands can be discerned in the relevant literature. The first, which can be termed 'institutionalist', and which continues to pervade research into French public policy-making, assumes that significant influences on the policy process tend to originate within the state – that is, with actors endowed with institutional, formally legitimated authority. The second, associated notably with the 'policy network' approach, takes a wider view of the influences which may come to bear on the elaboration and implementation of policy, and seeks to account, notably, for the role of extra-governmental interests. Both types of approach are relevant to our enquiry, although, as will quickly become apparent, the policy network approach holds out the prospect of a more comprehensive account of the factors shaping policy than an exclusive focus on the formal institutions of government.

2.2 INSTITUTIONALIST ACCOUNTS OF POLICY-MAKING: MODELS OF THE FRENCH CORE EXECUTIVE

We will base our discussion of 'institutionalist' approaches to the policy process on the varieties of 'core executive' goverment identified by Elgie (1991, 1993) in his analysis of prime ministerial influence in French public policy-making. Drawing on Dunleavy and Rhodes' overview of British 'core executive' studies (Dunleavy and Rhodes, 1990), Elgie (1993, p. 21) defines this complex of institutions as comprising, in the French context, the presidency, the Prime Minister and his or her office, government departments, executive committees, the central administration and diverse advisers and officials responsible to the government. Elgie identifies six 'models' – or, more accurately, views – of core executive operations in France, which he classes as 'presidential government', 'segmented decision-making', 'executive co-operation', 'prime ministerial government', 'ministerial government' and 'bureaucratic co-ordination'.

The first of the views identified by Elgie, 'presidential government', is defined as 'the exertion of monocratic authority by the President', or 'the situation where the President both initiates government policy and where he decides the content of such policy' (Elgie, 1993, p. 22). There is no differentiation, in this claim, between policy areas; a government's policy output in its entirety is seen as bearing the mark of the presidency's undivided sovereignty.

The second model, 'segmented decision-making', posits a sectoral division of responsibilities between the President and the Prime Minister. Elgie identifies two variants of this thesis – one of which assumes a traditional presidential *domaine réservé*, comprising foreign and defence policy, while the second variant, articulated notably by Wright (1989), asserts that the President's prerogative has increasingly been extended, such that the 'extended domain' is now made up of the traditional domain with the addition of economic, financial and industrial matters; social and environmental issues; 'politically delicate' or 'explosive' issues; and any other issues which happen to attract the personal interest of a particular President. As we have already noted, a variant of this thesis is frequently invoked in discussions of post-1959 cultural policy, which is assumed, either in its entirety or in part, to have moved progressively into the President's extended domain.

The third model identified by Elgie, 'executive co-operation', asserts that 'in all policy areas responsibility for policy-making is shared' by the President and the Prime Minister (Elgie, 1993, p. 28). Not only does the nature of the 1958 Constitution require 'routine administrative consultation' between the two; some proponents of the model also argue that 'for the smooth functioning of the system as a whole, the two parts of the executive have to work closely with each other'.

The fourth model, 'prime ministerial government', sees the Prime Minister as exercising undivided control over all important policy issues. Until 1986, this view represented little more than a purely juridical interpretation of the 1958 Constitution, with little practical relevance; however, it became more credible during the 1986–8 *cohabitation*, and should be seen as referring essentially to situations where the presidency and parliament have opposing political polarities.

According to the fifth model described by Elgie, 'ministerial government', 'the Minister is the chief policy-maker in his/her particular area and, when compared with the variants analyzed previously, the role of the President and the Prime Minister is much reduced' (Elgie, 1993, p. 32). In support of this claim, the model points, in particular, to the absence of a tradition of collective cabinet authority in France; to the weight of a minister's personal *cabinet* in dealings with the ministerial bureaucracy and with other parts of the core executive; to the high degree of administrative centralization within a ministry and not least, a minister's effective powers of appointment with regard to the heads of a ministry's *directions*; and to the importance of a ministry's territorial field services in policy implementation. Moreover, the ministerial government thesis is bolstered by the relative weakness of parliament, evident both in the virtual absence of parliamentary scrutiny of departmental decisions and in the ministerial prerogative of legislating by decree in areas delegated by parliament.

We can distinguish two varieties of ministerial government: first, a situation in which the minister, endowed with independent power resources, has effective policy-making autonomy *vis-à-vis* the other elements of the core executive, notably the President and Prime Minister; second, a situation where the minister's influence and status are dependent on the support of more senior actors. In the context of the Socialist governments of the 1980s, the latter version would seem to be more relevant, given that 'coalition' politics within the presidential bloc played a relatively minor role under Mitterrand, compared with, say, the Gaullist-centrist alliances of the late 1960s and 1970s, and that the power of individual ministers was widely perceived as issuing from the Elysée Palace. With regard to cultural policy, Rigaud (1986, p. 12) suggests that two conditions have to be met for a minister to be effective, both in terms of policy outputs and in terms of public visibility. First, 'the minister needs to be carried along by the circumstances and, more important still, by the unreserved support of the Prime Minister, the President, or both', and second, 'the minister needs to have under his responsibility a prominent policy area, in which a policy crucial to the government's programme is being applied, or which has priority in virtue of its deliberate originality. This has been the case of culture since 1981 . . . '. Ministerial autonomy, here, is severely circumscribed; however, there is still scope for effective policy formulation by the minister.

We should note the prominence of the ministerial government view both in general, non-academic accounts of cultural policy-making during the tenure of Jack Lang, and in historical accounts of the evolution of cultural policy since 1959, which is frequently subdivided into discrete ministerial 'eras', with responsibility for particular policy successes – or failures – being attributed to the personal qualities or enthusiasms of individual ministers.[4]

The sixth and final model discussed by Elgie, 'bureaucratic co-ordination', describes a situation where 'the political elements of the core executive have little or no control over the contents of policy decisions' and 'policy choices are determined by the bureaucratic elements of the core executive' (Elgie, 1993, p. 34). The bureaucratic co-ordination model obviously stands apart from the other views described, in its focus on non-elected decision-makers and in its affinity with 'elite theory' conceptions of the policy process. However, we might again observe briefly that the bureaucratic co-ordination view has a certain *prima facie* plausibility when applied to cultural policy, given the almost inevitable importance of unelected officials in the attribution of subsidies to individual institutions, particularly in the context of a system where the principle of 'peer assessment' plays only a subordinate role. We might also note that an extensive literature has been devoted to the supposed growth of a 'technocratic' policy-making elite under the Fifth Republic – the relevance of such conceptions to the policy area we are considering remains, of course, to be established on a case-by-case basis.

The differences between the six conceptions of the policy process set out above are substantial, and the views decribed by Elgie may constitute a useful framework for identifying the specificities of cultural policy-making in France. However, the claims made by the six core executive models have equally been worth summarizing for what they omit. Thus, while not all studies focusing on the core executive take care to set out their premises regarding the relationship between intra-governmental and 'societal' actors, there is an implicit assumption, in most cases, that the impact of extra-governmental actors on policy outputs is negligible or, at most, sporadic and accidental.[5]

While it will only be possible to assess the merits of the various models outlined here by testing them against the evidence obtained in a particular policy sector, there are legitimate grounds for *a priori* scepticism regarding the explanatory power of a purely institutionalist approach. First, with regard to the specific empirical issue we are considering, the policy 'paradox' we are concerned with – that is, the constraints on political 'voluntarism' manifest in the cultural policy of the French Socialists – could be perceived as casting doubt on the 'primacy of the political' implicit in institutionalist models, without our even having to look in detail at the mechanics of the policy process. Second, and more fundamentally, we may be motivated by a general 'bias' against conceptions of the social world which systematically exclude a vast range of extra-governmental actors from the focus of our research. Thus, Dunleavy and Rhodes (1990, pp. 19–20), in their survey of British core executive studies, point out the limitations of simple institutionalist accounts, and advocate 'a broader range of theories . . . to help create far more systematic and diversified modes of empirical analysis'; among these, they include 'differentiated accounts of the central executive', with 'the now widespread recognition of the importance of policy communities'.[6]

2.3 BEYOND INSTITUTIONALISM: POLICY NETWORKS, POLICY COMMUNITIES AND ISSUE NETWORKS

Against the institutionalist views adumbrated above, we might thus posit an approach to the study of policy-making which shifts the locus of analysis away from formal politico-administrative hierarchies, and which has increasingly found expression in the notion of 'policy networks' and 'policy communities'. The concerns which underlie the development of the policy network literature can be traced to Lowi's work on relations between government and interest groups in the United States (Lowi, 1964, 1969). The terminology of 'policy communities' as such originated in Heclo and Wildavsky's work on bureaucratic politics (Heclo and Wildavsky, 1974), and was later developed by Richardson and Jordan (1979), who argued, with regard to the policy process in British central government, that 'the policy-making map is in reality a series of vertical compartments or segments – each segment inhabited by a different set of organized groups and generally impenetrable by "unrecognized groups" or by the general public'; these segments are termed 'policy communities' (Richardson and Jordan, 1979, p. 174).

Indirectly, the development of the network concept was driven by scepticism regarding both 'the picture-book account of policy-making through party manifestos and Parliament' (Jordan, 1993, p. 58) and the assumptions of 'naïve' pluralist theory. At the same time, a number of normative concerns can be seen to lie behind much of the network literature: thus, Rhodes and Marsh (1992, p. 200), paraphrasing Lowi (1969), summarize the 'case against policy networks' by noting that they 'destroy political responsibility by shutting out the public; create privileged oligarchies; and are conservative in their impact because, for example, the rules of the game and access favour established interests.'

Progressively, the network concept has been refined to accommodate, and categorize, different varieties of the structures and processes identified. The core concept of 'policy community' has effectively been replaced by that of 'policy network'. An elementary definition of 'policy network' is offered by Rhodes (1985, p. 17), following Benson (1982, p. 148): a policy network is seen, here, as 'a complex of organisations connected to each other by resource dependencies and distinguished from other complexes by breaks in the structure of resource dependencies'. Policy communities, on the other hand, have increasingly come to be conceived of as being but one of several organizational types falling under this general definition. Rhodes and Marsh (1992, p. 186) thus define 'policy network' as 'the generic term encompassing all types'. Jordan and Schubert (1992) go further, suggesting that the network concept can replace existing theoretical models of state-group relations, having superseded the literature on different varieties of corporatism, 'group subgovernment' and 'iron triangles'. The central features of the policy network approach, for Jordan and Schubert, are the stress on 'interaction' and the low explanatory significance attributed to formal, constitutional distinctions between the state and societal actors; within this framework, a large variety of structures can of course be accommodated.

The specificity of 'policy communities', by contrast, is generally taken to lie in the closeness of relations between their members. The earliest work on policy communities, notably by Heclo and Wildavsky (1974), stressed the personal links between those involved in the formulation of policy. Rhodes (1985, p. 18) proposes a more abstract definition, suggesting that policy communities are distinguished by a high degree of integration and closure to outsiders. Elsewhere, Rhodes (1986) describes policy communities as representing one endpoint on a continuum of types of interest intermediation, the opposite extreme being constituted by 'issue networks', a term originally coined by Heclo (1978, p. 102). On this continuum, different types of network are differentiated according to the nature of their membership, the degree of integration and cohesion among network members, and the distribution of resources and balance of power between members of a network (see Table 2.1). Perhaps most important, and related to these dimensions, policy communities and issue networks are seen as diverging fundamentally in the extent to which their members have access to the decision-making process: according to Rhodes and Marsh (1992, p. 192), 'the issue network is conceptualized as the "outsider" against which [the policy community] defines itself. The issue network includes the excluded groups who have an interest upon the issue but have little or no access to government.'

Rhodes and Marsh (1992, p. 184) point out that the 'intermediate' cases in this table present difficulties for any attempt to create a neat ordering of types. While policy communities and issue networks can be seen to diverge clearly on the central dimension of 'integration', professional, intergovernmental and producer networks are hybrids, distinguished by the nature of the interests dominant within a network. Yet the type of interest which dominates a network may also be an issue in the case of policy communities. For this reason, Rhodes and Marsh suggest a revised framework, in which questions relating to the interests prevailing within a particular network are left open. The typology developed by Rhodes and Marsh (1992, p. 187; see

Table 2.1 Types of policy network: the Rhodes model

Type of network	*Characteristics*
Policy community	stable; highly restricted membership; vertical interdependence
Professional network	stable; highly restricted membership; vertical interdependence; serves interest of profession
Intergovernmental network	limited membership; limited vertical interdependence
Producer network	fluctuating membership; limited vertical interdependence; serves interest of producer
Issue network	unstable; large number of members; limited vertical interdependence

Source: adapted from Rhodes and Marsh (1992, p. 183).

Table 2.2 Rhodes and Marsh's typology of policy networks

Dimension	*Policy community*	*Issue network*
Membership	limited; some groups consciously excluded; economic and/or professional interests dominate	large; encompasses range of affected interests
Integration	frequent, high-quality interaction of all groups on all matters related to policy issue; membership, values and outcomes persistent over time; all participants share basic values and accept legitimacy of outcomes	contacts fluctuate in frequency and intensity; access fluctuates significantly; a measure of agreement exists but conflict is present
Resources	all participants have resources, basic relationship is an exchange relationship; participating organizations are hierarchical, leaders can deliver members	some participants may have resources, but they are limited; basic relationship is consultative; varied and variable distribution and capacity to regulate members within participating organizations
Power	balance of power between members; although one group may dominate, it must be a positive-sum game if community is to persist	unequal powers, reflects unequal resources and unequal access; zero-sum game

Source: adapted from Rhodes and Marsh (1992, p. 187).

Table 2.2), in focusing on ideal types, may provide a useful basis for characterizing and, above all, comparing networks both within and across policy sectors. We should not, of course, expect to find either of the two ideal types exemplified in full in a particular case.

The typology proposed by Rhodes and Marsh is only one of several competing attempts to coin a definitive usage for the different terms which have emerged in the network literature since the 1970s. An alternative typology has notably been proposed by Wilks and Wright (1987), who suggest that the terminology should be defined such as to capture not structural differences between networks, as in the Rhodes and Marsh framework, but, rather, different levels of policy-making. However, while there is much to be said for a disaggregation of policy areas into policy sectors, sub-sectors and issues, and a corresponding disaggregation of networks, as urged by Wilks and Wright, the continuum developed by Rhodes and Marsh has one decisive advantage. Unlike other typologies, it is geared explicitly to a comparative approach to the study of networks, and it identifies a small number of crucial dimensions which may serve as the basis for such a comparison, both within individual sectors and across sectors. Given that we will be pursuing a research strategy based on sectoral case studies, it can thus be assumed that the Rhodes and Marsh framework will be of greater utility to our analysis than competing approaches.

2.4 APPLYING THE NETWORK APPROACH TO CULTURAL POLICY

Although the use of the network metaphor in the analysis of public policy-making has gained wide currency within 'Anglo-Saxon' political science, the concept has rarely been applied to the study of the policy process in France. As Mény (1989, p. 355) points out, French political scientists have generally shown themselves to be indifferent *vis-à-vis* group politics, such that the contributions of British and American scholars continue to dominate the debate. This comparative academic neglect of state-group relations is, of course, closely related to a conception of state power which makes no allowance for the role of private interests – a conception which, however, bears little resemblance to the reality of interest intermediation in French politics. Thus, Mény notes, 'on the one hand, [the state] rejects [interest groups] in the name of its theoretical supremacy, while on the other hand, it establishes relations with them which are so close that it is sometimes difficult to discern the boundary between public and private interests, bodies and institutions' (Mény, 1989, p. 360). The hierarchical conception of the French polity spelt out by Mény, with its central image of a unitary *Etat* impervious to 'sectional' interests, has inevitably, and perhaps to an even greater extent than in other areas of public policy, coloured perceptions of the relationship between public agencies and their 'clients' in the domain of cultural policy. As in other policy areas, this conception is increasingly coming to appear inadequate – and the policy network approach could be seen as one way of introducing a greater degree of realism into debates about the role of the French state.

A potential problem arises, however, from the peculiar structural features of the policy area concerned. Much of the empirical work drawing on the network metaphor and, more important, the theoretical literature developed around the concept, focuses on either government-industry relations or the policy impact of groups involved in some form of public service delivery. There is, on the other hand, no precedent for the application of policy network analysis to cultural policy. Before attempting to transpose the concept to the area of arts funding, we should at least be aware of some of the differences with the policy areas more commonly studied.

Perhaps most important, economic interests are construed in much of the literature as being prior to, and in some sense independent of, the state or its component agencies; while the autonomy of the state may be at issue, the autonomy of business groups is usually taken as given – unless, that is, we are explicitly dealing with instances of 'state corporatism' or *pantouflage*.[7] Moreover, business groups will seek to extract economic or regulatory benefits from the state, and they will employ their resources to obtain these benefits. The relationship between cultural producer groups and the state presents an obvious anomaly here. With the exception of the 'cultural industries', the producers we are concerned with – that is, subsidized arts institutions – are dependent on public funding for their very survival, and consequently lack independent economic resources which they might use as 'levers' in their relations with state agencies.

In short, there are limits to the applicability of much of the literature on state-group relations to the area of cultural policy. However, even if there are, *prima facie*, strong dissimilarities between economic or industrial policy and arts funding, the models we have described may be useful analytical tools. Specifically, with regard to the question of 'autonomy', or its absence, in the case of subsidized institutions, financial dependence on the state cannot be taken as indicative by itself of a particular distribution of bargaining resources between an institution and a governmental agency. That is, even if we leave aside the crucial issue of the autonomy of the agency itself, the relationship between a subsidized institution and an agency may be influenced by the non-economic resources on which the institution is able to draw – that is, resources unrelated to the institution's actual economic significance. Specifically, these resources might involve the mobilization of organized professional or 'issue' groups; they might be derived from the technical expertise or prestige of such groups; or they might be a function of the social or economic status and influence of a particular group of cultural consumers.

A further important *prima facie* contrast between cultural and industrial policy arises from the 'structuring' effects of public funding policies on affected groups and organizations. As Saez (1985a, p. 408) points out in discussing state-group interactions in the implementation of *action culturelle*,

> each segment [of the bureaucracy] structures its clientèle while at the same time having to respond appropriately to its demands. Thus, the state apparatus is unable to give a single, harmonious and co-ordinated response to the problems of cultural development, but, to the contrary, has to individualize its reactions by helping to structure the interests of social groups and their representatives.

In addition to being dependent on the state for material resources, then, the organizational shape of 'client' groups may be determined by the nature of state intervention. However, the policy network approach is quite capable of accommodating the 'structuring' effects of an agency's policies. As Rhodes (1985, p. 14) points out, the structure of policy networks and their constituent group elements may be subject to intentional 'manipulation' by agencies of the state, and 'this manipulative capacity includes the ability to create and strengthen policy communities and not just to obtain preferred policy outcomes'. The extent of such 'manipulation' by the state remains a matter for empirical investigation, of course, and we will need to return to this issue in our sectoral case studies.

A related question concerns the direction of interaction within networks and the extent to which power – or, more specifically, the ability to influence policy outcomes – can be located within a network. Rhodes (1981, 1985) argues that networks should be seen as based on interdependence between actors, and suggests that the concept of 'power dependence' may be more appropriate to network-like relationships than the notion of 'power' as such: 'The importance of the concept of dependence, and the reason that it is stressed more than the concept of power, lies in the fact that it admits of reciprocity.' (Rhodes, 1981, p. 48) Van Waarden (1992, p. 31) similarly argues that 'networks do not necessarily have a power centre, and hence coordination is not by hierarchic authority (or, in more enlightened form, hierarchic consultation) but by horizontal bargaining.'

However, there is nothing in the network concept as such to exclude the possibility of resource or power differentials within a network. Indeed, the typology suggested by Rhodes and Marsh specifically accounts for this in the distinction between policy communities and issue networks, with various intermediate forms conceivable between the two ideal types. The points raised here regarding group 'autonomy' and the structuring impact of state policies on groups and networks – which are closely related to the issue of power relations inside networks – could moreover be expressed in terms of the 'resources' and 'power' dimensions in the Rhodes and Marsh typology. Again, only empirical analysis of the relations between different actors in the policy-making process on a sectoral basis will show what types of network prevailed in the cultural policy arena.

CONCLUSION

Existing accounts of French cultural policy-making, and attempts to elaborate theoretical models of cultural policy formation, point to the apparent 'insulation' of cultural policy outputs against explicitly political influences, and to the prevalence of a professional 'logic' in policy choices, as the central issue which any analysis of the cultural policy process has to address. However, they do not explain these phenomena; nor do any of the approaches to the study of cultural policy discussed in this chapter offer a satisfactory framework which we might apply in our six sectoral case studies. We noted that a suitable theoretical framework has to meet two conditions. First, it needs to account for the impact on policy outputs of relations between formal institutional actors – that is, the state – and the extra-governmental beneficiaries of public intervention. Second, it has to incorporate a comparative dimension, such as to enable meaningful conclusions to be drawn on the basis of a series of case studies.

To this end, the possible contribution to our analysis of a more general literature on public policy-making was considered. Of the various views and models discussed, the policy network approach appeared to offer the most promising theoretical basis for our analysis of French cultural policy-making. Specifically, we focused on the continuum of network types developed by Rhodes and Marsh, based on the distinction between 'policy communities' and 'issue networks'. Unlike narrowly institutionalist interpretations of the policy process, and to a greater extent than competing typologies of networks, the Rhodes and Marsh typology is capable both of accommodating extensive professional involvement in policy-making and of accounting for substantial variation in professional influence across networks and sectors. Above all, it is flexible enough not to pre-empt our empirical findings with regard to the internal structure of networks: both a statist pattern of policy-making, in which public actors dominate the policy process to the exclusion of societal interests, and more complex types of interaction between public and private actors are compatible with the framework proposed by Rhodes and Marsh. The actual balance between different actors within this framework remains to be ascertained empirically, of course: we thus turn to our first case study, dealing with one of the *grands projets* of the Mitterrand presidency, the *Cité de la Musique*.

3 The *Cité de la Musique*

The capital projects known as the *grands travaux* are perhaps the most ostentatious legacy of François Mitterrand's fourteen-year presidency. Yet while the *grands travaux* are widely seen as standing 'among the purest expressions of sovereign presidential power in the Fifth Republic' (Harrison, 1993, p. 210) and often taken as proof that cultural policy has, since Pompidou, moved into the presidential *domaine réservé*, the diversity of the projects concerned, and their rather dissimilar 'histories', nonetheless suggest a more complex reality. This chapter considers the policy process with regard to a single *grand projet*, the *Cité de la Musique*, the last phase of which was inaugurated in January 1995. The *Cité* comprises the new buildings of the Paris *Conservatoire*, a medium-sized concert hall, and a 'museum of music'.

The case study presented in this chapter should by no means be considered as aspiring to make claims about decision-making with regard to the *grands travaux* in general; however, it will question the extent to which it is possible to characterize the *grands travaux* without drawing on more discriminating concepts than a monolithic 'presidential government' thesis. The choice of the *Cité de la Musique* is not coincidental, nor is it entirely innocent given the argument which is pursued in this chapter. Perhaps more than the other cultural *grands travaux* initiated under the two Mitterrand presidencies, the *Cité de la Musique* project was closely bound up with existing sectoral policies, and necessarily entailed the mobilization of prospective private sector beneficiaries.

Two distinct, if related, issues will be addressed in this chapter. At one level, we consider the initial decision to build a 'grand ensemble musical', and the project's subsequent difficulties in the face of severe budgetary restrictions and the ambivalence of its political patrons. At another level, we attempt to explain the evolution of the project's content and configuration. As the *Cité*'s specifications continued to change frequently even after construction had begun, and as the influence of the main actors involved in the project's elaboration fluctuated considerably over time, these issues are examined concurrently, by considering the evolution of the project chronologically and attempting to isolate the actors responsible for specific decisions. Finally, we consider the implications of our findings; specifically, we ask whether the notion of presidentialism as applied to the *grands travaux* might usefully be tempered by according greater significance to the role played in the elaboration of a project such as the *Cité de la Musique* by interest groups and the bureaucracy – and by policy networks horizontally transcending the boundaries between the state and its 'clients'.

3.1 THE *CITÉ DE LA MUSIQUE*: ORIGINS, 1978–81

Though the *Cité de la Musique* was officially presented as one of François Mitterrand's *grands travaux* alongside such projects as the *Grand Louvre* and the *Très Grande Bibliothèque*, the possibility of constructing a musical *grand projet* of some sort at the Parc de la Villette, the site of a disused abbatoir in the north-east of Paris, was first mooted in the last years of the Giscard presidency. Giscard had approved the construction of a vast science and technology museum on the same site (later to be completed, with minor design changes, under Mitterrand), and a cultural project appeared desirable to complement the other elements of the development.[1] The planners' ambitions for the Villette site coincided with the demand, articulated with increasing urgency by leading representatives of the French musical profession, that the Paris *Conservatoire*, the most prestigious in France, be moved from its dilapidated site in central Paris and given a new home. It was thus assumed from a very early stage that any major new musical complex, be it at La Villette or at another site, should eventually accommodate the *Conservatoire National Supérieur de Musique et de Danse de Paris* (CNSMP). Yet while there was a large consensus among the musical profession about the need to create a new home for the CNSMP, it was far from obvious what the other elements of a 'musical city' should be. The project was thus prone from the very beginning to the competing demands of various institutions and individuals in the musical world. Foremost among these was Pierre Boulez, the single most influential figure in the French musical establishment, who, while taking a close interest in the plight of the *Conservatoire*, was also looking for a permanent base for his contemporary music ensemble, the *Ensemble InterContemporain* (EIC) – which, while generously subsidized by the Ministry, had only a temporary home at the *Cité Universitaire*. Other institutions were similarly tempted by the prospect of acquiring a share in the prospective 'musical city'. Thus, the *Orchestre de Paris* was looking to establish a permanent base in a suitable venue, its case strengthened by the fact that the capital lacked a properly equipped modern symphonic concert hall. A further, more distant project beginning to gain currency among some professionals – and, to a certain extent, among the politico-administrative elite – concerned a possible remedy to the limitations of the Paris state opera house, the nineteenth-century *Palais Garnier*.

A number of professionals, notably Boulez, pointing to the precedent of New York's Lincoln Center, argued that an 'integrated solution', combining a teaching institution (that is, the *Conservatoire*) and performance venues for both opera and symphonic music, had substantial intrinsic merit. However, in the actual decision-making process, the conception of the project as an integrated, coherent entity would rarely prevail – the project's form being determined essentially by the confluence of the desiderata expressed by various interested institutions and individuals. As we shall see, this was to prove crucial in conditioning the often disordered decision-making process once the definite construction of the *Cité* had been decided.

3.2 1981: THE *CITÉ* TAKES SHAPE

Following François Mitterrand's accession to the presidency and Jack Lang's appointment as Minister of Culture, the 'musical city' was included among the President's prospective *grands travaux*, more at Lang's insistence than because of any personal enthusiasm for the project on Mitterrand's part – in Lang's own words, 'François Mitterrand, at the beginning, was neither for nor against [the *Cité*]'.[2] In fact, most of the projects chosen in the wake of the Socialists' election victory (the final list of *grands travaux* planned for the first *septennat* was established by March 1982) responded to some previously identified planning need, or were simply variants of projects elaborated under the Giscard presidency, as in the case of the *Cité de la Musique*. Even the *Grande Arche de la Défense* and the *Grand Louvre*, the only *grands projets* which Mitterrand explicitly claimed as his own,[3] were in a sense developments of projects that had been initiated in the 1970s.

While the general principle of a musical *grand projet* was accepted by the Elysée at a very early stage, its precise shape emerged only gradually. As we have seen, the project that had been discussed for La Villette was in fact a conflation of several distinct initiatives – a new *Conservatoire*, a concert hall, a home for Boulez's EIC, and possibly even a new opera house. It quickly became apparent that the new Minister of Culture – who, like his predecessors, had soon established a close working relationship with Pierre Boulez – was committed to a major project of some description that took these very diverse demands into account: in August 1981, Lang spoke publicly of a 'musical Beaubourg' (that is, a musical equivalent to the *Centre Pompidou*), while Mitterrand, in his first press conference devoted to cultural issues, announced the construction of a 'cité internationale de la musique'.

However, neither Lang nor Mitterrand had very precise ideas about the actual shape such a project might assume. Lang's first concrete action regarding the *Cité*, in September 1981, therefore consisted in commissioning a report from the writer and diplomat Jean-Pierre Angremy, who was to assess the demand for new musical venues in Paris and suggest ways of incorporating these in a 'nouvel ensemble musical', of which the new opera house would form an integral part.[4] However, for Angremy the *Cité* as such was only of secondary importance: 'Of course, what really interests us about this project is the opera house. As for the rest, we will do our best, but we will do it well' (Rémy, 1989, p. 29).

Angremy submitted his report, which advocated a 'musical city' with a new *Conservatoire* and several auditoria, as well as an opera house, in December 1981. By this time, the choice of La Villette as the location of the projected complex no longer seemed obvious, as other contenders for the project had emerged. These included the more centrally located place de la Bastille, put forward by the Gaullist-controlled Paris city government, the *Ville de Paris*, as well as the high-rise office development of La Défense to the west of Paris and even the working-class suburb of Marne-la-Vallée. Moreover, a 'split' solution – separating the opera from the remainder of the project – increasingly gained currency among planners. This option appears to have emerged initially within the *Atelier Parisien d'Urbanisme* (APUR), the planning agency of the *Ville de Paris*, which suggested building the opera – the only element of the project which aroused any real enthusiasm on the part of planners and politicians – at the place de la Bastille, leaving the less prestigious elements, the *Cité* as such, to La Villette. Municipal planners, tacitly supported by the mayor, Jacques Chirac, were viscerally opposed to the *établissement public* of La Villette being made the 'gift' of yet another major capital project, not least since they were keen to recover as much as possible of the site for housing developments.[5]

Between October and December 1981, a number of inter-ministerial meetings were held in which the competing sites were presented to civil servants from the ministries concerned; the sites at La Villette and La Défense, in particular, were supported by strong bureaucratic lobbies linked to the *établissements publics* responsible for their development, while the APUR pushed very hard for the place de la Bastille, its main argument being the site's historical symbolism. On the other hand, the notion of building an opera house at Marne-la-Vallée, while characteristic of the ideological rhetoric that characterized the Socialists' first year in power and in keeping with the idea of a 'people's opera' publicly advanced by Mitterrand, had few serious advocates. The final decision about the project (or projects) lay with the President. Mitterrand's decision was made public in March 1982: the project would indeed be split in two, the *Cité* being built at La Villette and a separate competition being launched for a new opera house at the place de la Bastille. This solution, which diverged fundamentally from the 'integrated' project advocated by professionals like Pierre Boulez, would condition many of the problems that would subsequently bedevil both the *Cité* and the *Opéra Bastille*.

***Cité de la Musique* and *Opéra Bastille*: Presidentialism versus Cultural Policy?**

In attempting to explain the President's decision, it is difficult to discern a single rationale or, indeed, a single centre of decision-making. The go-ahead given to the 'core' of the *Cité* is perhaps least surprising, given the consensus among the profession, senior civil servants in the Ministry of Culture and key members of Jack Lang's *cabinet* about the need to re-house the *Conservatoire*. The decision to build a separate opera house at the Bastille, on the other hand, seems to have been motivated by reasons that had little to do with the actual content of the project, or indeed with cultural policy in a narrow sense. Thus, Mitterrand, who was known to have little interest in the opera project itself,[6] and his advisers seized on the plans for a new opera house as a project that could eventually be used as a symbolic centrepiece for the 1989 bicentenary celebrations. This preoccupation with what would essentially be a prestige object, unsuited to the functional and geographically marginal setting of La Villette, coincided with the concerns of some of Lang's advisers – notably the architect Christian Dupavillon, responsible for the *grands travaux* programme in Lang's *cabinet*, who favoured a strong architectural 'statement' at the Bastille, and who would later be responsible for organizing the bicentenary celebrations in which the *Opéra Bastille* figured prominently. The decision to site the opera at the Bastille was helped, furthermore, by the position of the *Ville de Paris* – which favoured a 'split' solution, and whose consent Mitterrand needed for a number of his more controversial projects. Indeed, the decision to separate the opera from the main project, publicly announced on 9 March 1982, was first leaked after Mitterrand had met Jacques Chirac at the Elysée on 11 February 1982 to discuss his *grands travaux* programme.

The decision to build a separate opera house at the Bastille was in a sense both the first and the last important decision taken by Mitterrand with regard to the *Cité*, but one that would bear heavily on the evolution of the project. In the words of Boulez, who had been the most prominent advocate of the 'integrated' *Cité*, and whose advice had, uncharacteristically for a decision with important implications for music policy, been ignored,

> [when] the project was launched, [and] the opera house was put at the Bastille, all of a sudden, the Cité de la Musique was called into question. Because ultimately, when one only has an institution like the Conservatoire, everyone says, 'Of course, it has to be done' – but it isn't prestigious.... And so La Villette was never considered anything but secondary.[7]

It was clear from the outset, then, that the Bastille project – not least since it had to be completed by 1989 – would be given priority in the Ministry's spending plans, and receive much closer political attention. This had two implications for the *Cité*: first, obtaining an adequate budget for the project was destined to be problematic; second, the absence of clear presidential or ministerial leadership would give extra-governmental actors – and more specifically, competing elements within the musical profession – an important role in the project's elaboration.

3.3 1982–4: A DIFFICULT GESTATION

Once it had been decided in principle to build a *Cité de la Musique* at La Villette, the crucial issue of who, or which institution, would be given the role of elaborating and then implementing a construction programme became more acute. The administrative structure traditionally chosen for the construction of presidential *grands projets* was that of an administrative *établissement public*, placed under the authority of the ministry nominally responsible for the project concerned, but with a President directly answerable to the President of the Republic. The Parc de la Villette, however, into which the *Cité de la Musique* would be inserted, grouped a number of existing projects under the authority of a single *Etablissement Public du Parc de la Villette* (EPPV), attached not to the Ministry of Culture but to the Ministry of Public Works.[8] Rather than create a separate legal structure to oversee construction of the *Cité*, it was decided that the EPPV would act as *établissement constructeur* – that is, while the Ministry of Culture provided the funds, they would be dispensed *via* the EPPV, which thus acquired a substantial role in the execution of the architectural programme. On the other hand, the task of defining the 'content' of the Cité – notably by elaborating a programme for the architectural competition – was devolved to a specialized working group of civil servants and architects attached to the EPPV, the *mission musique*. The *mission* was created in July 1982, with an *énarque* and former administrator of the *Orchestre de Paris*, Jean-Pierre Guillard, appointed as its first director.

The curious administrative status of the *mission musique* merits closer attention, as it is indicative of the ambiguity which surrounded decision-making concerning the *Cité* in general. Though formally integrated into the EPPV, the *mission musique* would be answerable, where artistic matters were concerned, to the music directorate of the Ministry of Culture, the *direction de la Musique et de la Danse* (DMD). In fact, in yet another deviation from a simple hierarchical conception of decision-making with respect to the *grands travaux*, both the EPPV and its *mission musique* would enjoy a considerable degree of autonomy. In the words of one former DMD official,

> the dual influence of the President and of the Ministry of Culture in fact gave the *établissement public* charged with bulding the project, and its *mission* of experts, real autonomy. And consequently, a huge amount of room to manoeuvre for people who found themselves halfway between the Elysée Palace and the Ministry of Culture, and who were able to play out the one against the other – or, at times, to play the one without the other..... That is, there might be a discussion between La Villette and the Elysée, and this might give rise to a number of conclusions, and then these conclusions were announced to the Ministry of Culture as decisions which has been taken definitively – and vice versa. And to a certain extent that was, in the beginning, the game played by Jean-Pierre Guillard...[9]

Specifically, Guillard had the task of preparing the brief for the competition that would designate the *Cité*'s architect. In this, the *mission musique* was given almost complete freedom, working on the very general premise that the *Cité* would comprise a new *Conservatoire*, three different-sized concert halls, and a museum to house the *Conservatoire*'s collection of musical instruments. The Ministry's *direction de la Musique*, responsible, in theory, for exercising artistic control over the project, remained detached from the *mission*'s day-to-day work,[10] and the *mission* effectively had much closer relations with the EPPV and its President, Paul Delouvrier, a high-ranking civil servant who had considerable weight in his own right in his dealings with the Elysée and the government.[11]

The *mission*'s autonomy in preparing the competition was further strengthened by the virtual absence of the *Cité*'s future 'clients' in its deliberations at this stage. The one notable exception was Pierre Boulez, who had been involved in the project's elaboration throughout. The only major decision on which the composer was not consulted, according to Boulez himself, was the initial decision to split the *Cité* and the new opera house. In 1983, on Guillard's initiative, Jack Lang promised Boulez that the EIC would definitely find a permanent home at the *Cité*, and that it would be the main user of one of the planned concert halls. As Boulez recalls,

> there were meetings in Lang's office, with Guillard, and at that point in time, the idea was definitely that the concert hall would officially be shared – because one had to maintain the fictitious idea of something belonging to the Conservatoire – between the Conservatoire and the *Ensemble InterContemporain*.[12]

The establishment of a clear linkage between the EIC and a specific part of the project changed the nature of Boulez's involvement in the project, which became much more focused. Thus, Boulez became closely involved in elaborating the technical specifications of the medium-sized concert hall which was to become the EIC's base. Virtually all important decisions taken by the *mission musique* or, at a higher level, within the Ministry, with regard to this part of the project were directly attributable to Boulez. Thus, it was largely at Boulez's insistence that the *mission musique* integrated the notion of 'modulability' into the hall's specifications – that is, the possibility of changing the hall's scenic layout to accommodate different spatial constellations according to the work performed.[13] Officially, it was still assumed at this stage that the *salle modulable* would eventually come under the control of the *Conservatoire*, which would share its use with the EIC. In fact, Boulez, with the tacit consent of the *mission musique*, the DMD and the Minister, came to acquire a dominant influence over decision-making with regard to the *salle modulable*, which would privately be referred to by Ministry officials as 'Boulez's concert hall'. As we shall see, this would have important consequences for the design and cost of the hall, for the overall shape of the *Cité*, and for the project's fate on the different occasions on which it was subsequently challenged for political or budgetary reasons.

Boulez's role was strengthened by the virtual absence, at this stage, of the *Cité*'s other 'clients' in the decision-making process. The then director of the *Conservatoire*,[14] a traditionalist opposed to the project, had virtually no contact with the *mission musique*. Yet in view of the political and professional consensus about the need to move the *Conservatoire*, which had been the project's original *raison d'être*, this part of the *Cité*, like Boulez's *salle modulable*, was not in doubt. By contrast, other elements of the original project, which lacked the same degree of support from influential extra-governmental actors, came to look increasingly doubtful. The symphonic concert hall, in particular, had no obvious client. Implicitly, it was assumed (not least since Guillard had been its administrator) that the *Orchestre de Paris* would eventually take up residence at the *Cité*; however, the orchestra had moved only in October 1981 to a refurbished nineteenth-century hall in central Paris, the *Salle Pleyel*, and did not push for a concert hall at the *Cité* with great urgency. Its position *vis-à-vis* the *Cité* was further complicated by the fact that it received one half of its funding from the *Ville de Paris*, which meant that any stake it might have in the project would inevitably be tied up with the complex relationship between the Socialist government and the mayor of Paris.[15] Moreover, the ever-greater role played by Boulez meant that the focus of the original project shifted somewhat: according to one senior official at the *mission musique*,

> what happened is that Boulez came and said, 'Stop! The *Orchestre de Paris* is nothing to do with me – it is I that do not have a concert hall, it is the Ensemble InterContemporain!' ... Boulez said to himself, 'I'd never get what I want, I'd only be second-in-line next to a large concert hall for the Orchestre de Paris.'[16]

Crucially, around the same time, it became clear that the *Cité*'s prospective budget would be nowhere near as substantial as initially hoped. The government's overall budgetary outlook had changed dramatically since the expansionary optimism of 1981; moreover, there was no doubt that priority would be given to the *Opéra Bastille*.[17] By mid-1983, it had become apparent that the project in its entirety would be unsustainable; it was almost inevitable, then, that the element least strongly articulated at the political level – in other words, the symphonic auditorium – would have to cede in the face of increasing budgetary pressures, along with a number of smaller components, such as a smaller 800–seat concert hall and a 'community hall' initially sketched in the programme. In fact, the decision to eliminate the large auditorium was never formally taken: officially, the project was merely 'phased', the second 'phase', comprising the large auditorium, being postponed indefinitely. However, as Marc Bleuse, Guillard's successor as director of the *mission musique*, points out,

> The 'nyet' for the concert halls came at the beginning of 1983. They didn't last. ... [Lang] saw the dossier, he saw the costs involved, he said, 'In any case, it isn't worth it.' He understood very quickly, and very quickly, he said to himself, 'On the other hand, we're going to ask Mitterrand to push for an opera house – which will be a much more distinctive project, at any rate it will really be something more personal – instead of the concert halls ...'[18]

To complement this gradual shift in the project's focus, the composition of the *mission musique* itself underwent significant change. Thus, Guillard, judged to have made insufficient progress on elaborating a programme for the prospective architectural competition, was effectively dismissed in early 1984 and replaced by the EPPV's President, Paul Delouvrier, with the composer Marc Bleuse, until then *inspecteur général*[19] at the *direction de la Musique*.[20] Maurice Fleuret, *directeur de la Musique*, had only just appointed Bleuse as new director of the *Conservatoire*; his simultaneous appointment as director of the *mission musique* thus corresponded to a desire on the part of both Fleuret and Delouvrier to refocus the project and give much greater weight and coherence to the new *Conservatoire*. The *Orchestre de Paris* 'option', by contrast, was now definitely discarded.

1982–3: a Network is Born

It may be helpful to summarize briefly the constellation of actors who would, with variations, determine the *Cité*'s subsequent development, and who were essentially in place by late 1983. The *Cité* 'network' that began to emerge at this stage was constituted around the officials of the *mission musique*, who possessed a large measure of autonomy *vis-à-vis* the Ministry of Culture in virtue of the *mission*'s attachment to the EPPV and its ambiguous position between the Ministry and the presidency of the Republic; the latter being remarkably absent from deliberations about the project once the fateful decision to split the *Cité* and the new opera house had been taken. On the other hand, the *mission* worked closely with one of its putative 'clients', the EIC, and its President, Pierre Boulez, to the extent of virtually deferring to Boulez in the drafting of the specifications for one important element of the *Cité*, while consolidating Boulez's support – crucial at the political level – for the overall project. Other potential 'clients', by contrast, had at best a distant relationship with the *mission* – the CNSMP remained an outsider to the network until the appointment of Marc Bleuse; the *Orchestre de Paris* was ambivalent in its commitment to the planned symphony hall and lacked, in any case, the political weight of Pierre Boulez. As we shall see, this cluster of actors, and its correspondence (or lack thereof) with the architectural programme of the *Cité* would, more than the decisions of the Minister of Culture or any other element of the political core executive, shape the future *Cité*.

3.4 1984–6: LAYING THE GROUNDWORK

The Architectural Competition

By January 1984, the *mission musique* had prepared a detailed programme for the *Cité*. The process by which the architect of the *Cité* was chosen, on the basis of this programme, is revelatory, again, of the nature of political, and especially presidential, involvement in the *grands travaux*. The procedure chosen for the competiton had two distinctive features. First, in contrast with the open, international competition that designated the architect of the *Opéra Bastille*, it was decided that the competition for the *Cité* would be open only to a specially invited group of French architects. Second, the competition would be held over two rounds: in the first round of the competition, candidates would be asked to submit an entry for the whole project – that is, for both 'phases'; a jury would then forward a shortlist of entries for the second round to the President of the Republic. Formally, the jury's role was only advisory and the final decision rested with the President. In the second round, the shortlisted candidates would be asked to submit a more detailed entry. While these would be considered once more by the same jury, the final decision lay, again, with the President.

The list of architects to be invited for the first round had initially been drawn up by the officials of the *mission musique*. However, as one former EPPV official observes, the process was characterized by political interference at the *cabinet* level from a very early stage:

> So, we had all these lists of architects, and we'd looked at their work, and we had portfolios, we had this and that. But actually, the list of architects who'd be invited came out of the pockets of a certain man in the Ministry of Culture. ... It was quite notable for the absence of certain architects, and the inclusion of others.[21]

On the basis of this list, the jury, meeting for the first round from 2 to 5 May 1984, then selected six projects, in accordance with the rules of the competition, which it suggested should go through to the second round. At the same time, it underlined its particular preference for two entrants, who received an equal number of votes: Henri Gaudin and Christian de Portzamparc. The names of the six candidates pre-selected by the jury were published in mid-June. However, to the dismay not only of the jury itself, but also of the French architectural profession at large, the President decided to ignore the jury's recommendations: only three projects were designated for the second round, de Portzamparc's among them. Henri Gaudin, on the other hand, had been struck off the shortlist. The President's decision provoked the resignation of all the architects on the jury; the *mission musique* thus had to hastily assemble another jury. The second round of the competition was held on 22 and 23 November 1984, and Mitterrand's final decision was made known in January 1985. The President had chosen de Portzamparc. However, it would be wrong to attribute Mitterrand's decision to override the jury simply to a desire to assert the 'fait du prince'. Thus, it was widely suspected that prominent members of Lang's *cabinet* – who were frequently consulted by Mitterrand about his architectural choices – had a strong preference for de Portzamparc's entry, not least because his entry for the *Opéra Bastille* contest, generally considered superior to the winning design, had been disqualified on technical grounds. According to one former EPPV official, 'it became clear during the technical analysis [for the second round] that the project was earmarked for de Portzamparc'[22] – specifically, the *mission musique* was said to have arranged the costings for the different projects such as to give an advantage to de Portzamparc's entry, which, given its complex layout, was in fact much more costly than its competitors.

Once the project's architect had been designated, the *mission musique* could pursue its task of developing a more detailed technical brief.

Concomitantly, the core 'network' of decision-makers involved with the *Cité* was consolidated. The priority accorded the *Conservatoire* was confirmed, in virtue of Bleuse's dual function as head of the *mission* and the CNSMP. Moreover, certain members of the CNSMP teaching staff began to play a more active role in elaborating the technical specifications not only of the *Conservatoire*, but also of the planned musical museum, both unofficially and as consultants to the *mission musique*, emphasizing the extent to which the distinction between the *mission* and its 'clients' had become blurred. With regard to the design of the *salle modulable* – the core of the *Cité* proper – Boulez's close collaboration with the *mission musique*, and now the architect, intensified, and the technical specifications suggested by Boulez became more precise. Thus, in addition to his initial demand of spatial variability, Boulez insisted that the hall have variable acoustics; both of these demands (which clearly reflected the priorities of the EIC, and promised little benefit for the CNSMP, to which the hall would, theoretically, be attached) were accepted by the *mission musique*, as was the allocation of generous office space to the EIC.

Thus, the period immediately after the architectural competition saw the emergence of a pattern of intense co-operation between the *mission musique*, the *Cité*'s two major 'clients', and, to a lesser extent, senior civil servants following the project at the *direction de la Musique*. On the other hand, the close identification of the *mission musique* with its 'clients' made for a tense relationship with the EPPV and Serge Goldberg, Delouvrier's successor as President, who was responsible for the actual construction of the *Cité*. In the words of one former *mission* official,

> It was Goldberg's mission to get the project built at any cost, and so the 'users' were seen to a certain degree as being in the way . . . . And one was caught between three lines of fire – there was the *établissement public* responsible for construction, there was de Portzamparc, who was on our side – or rather, on the side of the end-users – and there was us. And this very deep understanding between the users and the architect profoundly annoyed the *établissement public*, because in addition there was a blurring of roles between the *mission musique* and the users.[23]

This constellation of interests, and the ensuing potential for conflict, would be intensified by the financial uncertainty with which the *Cité*'s planners had to contend at all stages of its design and construction.

The Budget of the *Cité de la Musique*: an 'Incrementalist' *Grand Projet*?

The *Cité de la Musique* that emerged towards the end of the architectural competition was a drastically reduced version of the 'grand ensemble musical' that had been planned initially. Nonetheless, an adequate budget was by no means assured even for the trimmed version of the *Cité*. The *Cité* was strongly supported by the musical profession and its allies in the *direction de la Musique*, led by Maurice Fleuret; at the political level, by contrast, the *Cité* was widely seen, and most crucially by members of Lang's *cabinet* and the Elysée staff, as a poor relation of the more glamorous Bastille project, which never suffered from the same ambiguity as to its 'content', and which had a definite construction timetable. To compound the project's difficulties, its prolonged gestation period meant that it was handicapped, in budgetary terms, *vis-à-vis* those *grands projets* which had been initiated at the very beginning of the first Mitterrand *septennat*. Credits allocated to the *grands travaux* overall had peaked in 1984, at 5.3 billion francs, falling to 3.9 billion francs in 1985 and 2.8 billion francs in 1986 (of which 1.1 billion francs was budgeted for the *Opéra Bastille* alone).[24] Moreover, the budget of the Ministry of Culture, almost doubled in the wake of the Socialists' 1981 election victory, had ceased to expand, and had in fact declined by around 10 per cent from 1983 to 1984.[25] Against this background, funding for the *Cité* would inevitably be tight.

Initially, in 1984, a sum of around 400 million francs was budgeted for the two 'halves' of the *Cité* (that is, the *Conservatoire* and the concert hall and museum, which came to be known as 'partie Ouest' and 'partie Est' because of their geographical location on the Villette site); and it was on this basis that de Portzamparc signed a contract for the design and construction of the *Cité*. In fact, it was an open secret among those involved in the project's elaboration that the budget had been deliberately and grossly underestimated from the very beginning, and that each of the two halves of the *Cité* would cost at least 400 million francs. Thus, as Marc Bleuse recalls,

> When I was appointed by Delouvrier, he told me, 'We've been given 400 million francs to build both halves of the *Cité de la Musique*, West and East. Tell everyone that the figures are incorrect, that we need twice as much. Because one day people will blame you, they will tell you, "You're the Director of the *mission musique*, you're in charge of this project, you didn't realize..."'[26]

As Delouvrier had predicted, budget estimates for the *Cité* would frequently be revised, to more than 600 million francs in November 1985 and 800 million francs in 1990; the final cost of the two parts of the *Cité* would attain around 1.3 billion francs. Indeed, in the words of Emile Biasini, appointed *secrétaire d'Etat* for the *grands travaux* by Mitterrand in 1988,

> the Minister never asked for a [definite] budgetary envelope, because he had never really told the President, 'I'm going to build the *Cité de la Musique*' – assuming that, in any case, he would obtain the necessary funding on a year-by-year basis. When I took over, the project survived on the annual allocation of funds by the Budget Ministry.... So people would ask for 100 million, then 200 million, every time a new problem came up – when they had difficulties with the contractors, or when the digging didn't go as planned.... They were out of their depth.[27]

Thus, for want of a realistic budget set at the outset, the project survived only on increments periodically obtained by Jack Lang, supported by direct lobbying of the *direction du Budget* by the EPPV. As Lang recalls, the *Cité* received little direct support in the budget process from the Elysée Palace, and '[Mitterrand's] advisers who dealt with the Budget Ministry didn't care much about the project'.[28] Given what Lang terms the general 'indifference' facing the project, the support of Pierre Boulez became a crucial argument *vis-à-vis* both the President and the Ministry of Finance. Indeed, on several occasions from 1982 onwards, Lang asked the composer to write directly to the President ahead of important budgetary decisions, or *arbitrages*:

> On several occasions, the project nearly ran aground at the time of the budgetary *arbitrages*, because no one was interested in it. And more than once, I asked Boulez to write to François Mitterrand – which he did, on at least two separate occasions, in crucial periods during the preparation of the state budget, in June. And François Mitterrand was indeed influenced because Boulez was there. Without him, perhaps this project would not exist.[29]

However, the financial uncertainty with which the project was afflicted inevitably rendered the policy process discontinuous and unpredictable. It also had important implications for the relationship between the project's political backers and the civil servants responsible for its implementation. Thus, according to a former official of the *direction de la Musique*, the incessant budgetary difficulties gave a position of strength to the *établissement public* responsible for construction, in the sense that the *établissement public* – and no one else – was able to control the costs of the project. In other words, if at any time the Elysée Palace said, something needs to be done like this, or like that, the *établissement public* could reply, 'Well no, that's impossible, otherwise we're going to exceed the spending limits.' And the same went for the Ministry of Culture.[30]

A similar conjunction of incremental budgeting and extensive control of the programme by the 'implementing' bureaucracy prevailed during the *cohabitation* of 1986–8, which, paradoxically, saw a further weakening of hierarchical political control of the project.

3.5 1986: 'HIGH' POLITICS INTRUDES ON THE *CITÉ*

Following the defeat of the Socialists in the legislative elections of March 1986, Jack Lang was replaced by François Léotard as Minister of Culture. Briefly, the entire *grands travaux* programme, associated by the Right with Socialist profligacy and megalomania, seemed threatened. The new Prime Minister, Jacques Chirac, who ultimately controlled the budgetary purse-strings, was at best indifferent to the *grands travaux* – except in so far as they touched on his immediate interests as mayor of Paris. The attitude Léotard himself, who had been awarded the Culture portfolio only as a consolation prize, would adopt, was impossible to predict. In fact, the new government's position was anything but predetermined, and the fate of the *Cité* would be resolved only after protracted and often confused negotiations between different members of the government, on the one hand, and between the government and the professionals and bureaucrats involved in the project, on the other. The role of the President was, in this period, marginal.

Léotard's first decision with regard to the *Cité de la Musique* was the appointment of a commission, headed by the businessman Olivier Chevrillon, that would consider the future of both the *Cité* and the *Opéra Bastille*. The unstated brief of the *mission de réflexion* appointed in April 1986 was to find ways of cutting the overall cost of the Bastille and Villette projects.[31] More subtly, however, it also reflected the antipathy of important elements of the conservative establishment, strongly attached to the *Palais Garnier*, with regard to the *Opéra Bastille* – in fact, the commission's vice-president, the *énarque* and journalist Maryvonne de Saint-Pulgent, had been one of the project's most vociferous public critics. In early June, the commission submitted its confidential proposals to Léotard and Chirac: rather than pursue the construction of two separate projects, the Chevrillon Report advocated combining a truncated version of the partly completed *Opéra Bastille* and the main components of the *Cité* on the Bastille site – concomitantly preserving the traditional vocation of the *Palais Garnier*.[32]

Once the gist of Chevrillon's proposals had become known, and it seemed possible that the *Cité* would be scrapped before construction had even begun, the project's main advocates started to lobby the new Minister and his *cabinet*. Pierre Boulez was horrified at the prospect of seeing both the *Cité* and the new opera project diluted or scrapped. Consequently, Boulez – who, while on good personal terms with Lang, had always stood aloof from party politics – intervened both with Léotard directly and with a second commission appointed by Léotard in October 1986, the 'Commission Jacob', which was to make more precise recommendations for the Bastille site. According to Boulez,

> first of all, we 'demolished' the replacement project intended for the Bastille, that was ultimately the tactics used. I plucked bare the replacement project. So that's how we saved La Villette, by averting this ludicrous idea which consisted in moving the Villette [project] to the Bastille. And [we] also saved the [Opéra] Bastille by putting down all the projects they were putting in the place of what they wanted to get rid of.[33]

At the same time, though independently of Boulez's efforts, Marc Bleuse, who had remained in his post as director of the CNSMP and *mission musique*, intervened with his political superiors. The turning point for the future of the new *Conservatoire*, according to Bleuse, was a meeting with Léotard, requested by Bleuse, after rumours about Chevrillon's recommendations had filtered through to the *mission musique*. Bleuse's account of the meeting – confirmed by other sources – is worth quoting at length:

> I arrived in Léotard's office, carrying my project, and I must have performed an incredible act for him, with the model [of the *Cité*]. I explained everything to him, the entire background, everything we had done, everything we wanted to do, who I was, why I was doing it – and he was fantastic. After one hour – and it's rare for a minister to listen to you for one hour, without moving, with two people next to him – he told me, 'Monsieur Bleuse, you have won. In any case, I'm

> now entirely on your side. Everything I've been told up to now is nonsense, I'm going to get rid of this commission, and we're going to see the Minister of Finance together..., to make sure we're not given any trouble.'[34]

Chirac, however, who had also received Chevrillon's report, and who was initially inclined to support its conclusions, still had to be won over. Thus, Bleuse did in fact accompany Léotard to an 'audience' with the then Minister of Finance, Edouard Balladur; finally, Léotard intervened, successfully, with Chirac to obtain a green light for the project.

Bleuse's personal initiative, which undoubtedly played a crucial role in saving the *Cité*, was helped by two less personal factors. First, the need to create a new home for the CNSMP was widely recognized on both sides of the political spectrum. Thus, in the words of one senior official at the *direction de la Musique*,

> the concert hall, the cost of the project, and so on, did indeed come in for criticism. On the other hand, the transfer of the *Conservatoire* was not criticized at all. ... The politicians understood very well that even if it wasn't a priority for them, calling into question the transfer of the *Conservatoire* from the rue de Madrid would have provoked an outcry – including in right-wing circles. Because much as the Right lacked its own networks in the world of professional musicians, it is very well represented on the teaching side.[35]

Second, Léotard had good party political reasons for supporting the threatened projects: against the background of a continuous tug-of-war between the different components of the governmental alliance, Léotard, as a leading member of the *Parti républicain*, could scarce afford to be seen to concede too much ground to the Gaullist Prime Minister and Minister of Finance. Moreover, Léotard was anxious not to begin his tenure at the Ministry simply by terminating projects initiated by Jack Lang – it was clear that the performance of any incoming Minister of Culture would inevitably be judged first and foremost by comparison with his immensely popular predecessor.[36]

The *Cité* and Cohabitation

By the autumn of 1986, the project's future seemed assured, and construction work could begin. However, the financial difficulties of the *Cité* were by no means resolved. None of the successive budgetary extensions obtained by the EPPV was sufficient to maintain the original project in its entirety, such that the *Conservatoire*, in particular, had to be pared back substantially. However, at no point was a realistic overall budget put forward, and the programme itself continued to change frequently as a result of both the budgetary uncertainty and the evolving demands of the project's 'clients'.

The lack of coherence in the policy-making process was exacerbated, throughout this period, by the absence of a clear centre of decision-making: while the link between the *Conservatoire* and the *mission musique* had, through the dual directorship of Marc Bleuse, provided a centre of gravity for the project, the elaboration of the *Cité*'s programme became more disaggregated in the autumn of 1986, when Bleuse assumed the post of *directeur de la Musique* in the Ministry of Culture to replace Maurice Fleuret.[37] While having to relinquish his post as director of the CNSMP, Bleuse retained his position at the *mission musique*. As a consequence, the DMD's role as the administrative authority responsible for the project was further diluted – indeed, in the words of one senior *mission* official, 'the *établissement public* did not in fact obey any authority, and no one dared exercise any authority over the *établissement public*, neither the Ministry of Public Works nor the Ministry of Culture. In fact, the statutes were drawn up such that the line of authority was appallingly muddled.'[38] This facilitated the direct access of the *Cité*'s 'clients' to the *mission musique* and the EPPV, and left the field to an often acrimonious struggle for influence among the bureaucrats and professionals involved in the project, both at La Villette and at the Ministry of Culture. At the same time, Bleuse's move to the DMD left the *mission musique* effectively rudderless, and weakened the *Conservatoire* as an institution. The new director of the CNSMP, Alain Louvier, lacked Bleuse's authority *vis-à-vis* both the *mission* and the EPPV. What influence Bleuse and the *mission musique* lost in this period was largely recovered by the EPPV and its President, who effectively acted as an arbiter between the demands expressed by the prospective 'clients', and who, in virtue of his control over the project's budget, played a central role in relaying the *a posteriori* cuts made inevitable by the gap between the initial estimates and the project's actual cost.

3.6 1988–95: A TROUBLED FINALE

The return, in 1988, of the Socialists to government and of Jack Lang to the Ministry of Culture did not, as might have been expected, clarify the complex relationship between the different actors involved in the project. Nor did it lay to rest the ambiguity that surrounded the final design and future status of the 'partie Est' – the completion of which, unlike the *Conservatoire* buildings, which were inaugurated in 1990, still seemed a distant prospect. However, following Lang's return, a number of 'exogenous' events impacted in varying degrees on the project, though without modifying entirely the diffuse, and often confused, policy-making style that had characterized the project in the past.

First, in order to resolve the doubts surrounding the project, Lang, following a long-standing tradition for incoming ministers, commissioned a report on the project's advancement from Maurice Fleuret – who, in the words of one DMD source, 'was horrified at the way in which the project had gone adrift over a period of two years'.[39] Fleuret's report, which was never made public but which was understood to emphasize that the project lacked effective political or administrative supervision and was highly critical of the decisions made by the EPPV, to which it imputed serious financial irregularities, was seen by some officials as 'the first stage of his seizure of power'[40] at the *Cité*. However, Fleuret died soon after the report was submitted to Lang in 1989 – consequently, the impact of the 'rapport Fleuret' remained limited.

Perhaps more important than Fleuret's 'audit' in reminding policy-makers that the handling of the *Cité* needed to be reconsidered was François Mitterrand's decision, after his re-election in 1988, to appoint Emile Biasini, who had been responsible for the *Grand Louvre* project, to the newly created post of *secrétaire d'Etat*, or junior minister, responsible for the *grands travaux*.[41] Biasini, nominally attached to the Ministry of Culture but ultimately answerable to the President alone, had the brief, in stark contrast with the concerns of all other actors, of completing the outstanding *grands projets* on time and within budget. Faced with the uncertainties still surrounding the programme for the 'partie Est' of the Cité, Biasini privately threatened simply to halt the project. In Biasini's own words,

> at the *Cité de la Musique*, there was no programme. They didn't know what they wanted to do. So I told Portzamparc and the others, 'Either you tell me what you want to do, we establish a budget and I defend it, or else we stop the project altogether.'[42]

Again, as in 1986, Boulez played a central role in saving the project, by intervening directly with Biasini.[43] Biasini lifted his threat, and the eventual completion of the 'partie Est' now seemed more or less guaranteed.

Biasini's appointment, and his role with regard to the *Cité*, might at first sight be interpreted as a reassertion of presidential authority; a notion that could be seen to be supported by Biasini's deeply conflictual relationship with Lang. In fact, while Biasini pushed very hard for setting a definite programme, budget and completion date, and while this frequently brought him into conflict with Lang, the new *secrétaire d'Etat* was in no way involved in shaping the project as such – which, in any case, he viewed with profound scepticism. Nor, according to Biasini, was the President at all interested in bringing his weight to bear on the project:

> Given that he had accepted the inclusion of a *Cité de la Musique* in his programme, the President of the Republic, as a matter of principle, expected that a *Cité de la Musique* be built. But how, and what would be inside it? [His position was:] 'I've accepted the *Cité de la Musique*, I want the *Cité de la Musique*, now sort things out yourselves.'[44]

Thus, the President's detachment with regard to the project remained almost complete. By contrast, the influence of the *Cité*'s clients, and notably of Pierre Boulez, became progressively more pronounced.

1989–95: *'L'apothéose de Boulez'*

The single most important factor determining the evolution of the *Cité de la Musique* in its final phase was not the stance of the President, but the role played by Pierre Boulez, anxious to protect the part of the *Cité* which most directly concerned him, the *salle modulable*. Alarmed at the fate a similar brainchild of his, the projected *salle modulable* of the *Opéra Bastille*, had suffered as a consequence of a politico-administrative imbroglio not dissimilar to that which prevailed at La Villette, Boulez once again came to follow the project very closely, and used his influence to bear heavily on all important decisions relating to the *Cité* in its final phase. Boulez's role during this period has to be seen against the background of his conflict with Michel Schneider, who had been appointed *directeur de la Musique* at the Ministry of Culture in 1988 to replace Marc Bleuse, and who openly challenged what he saw as Boulez's hegemony in the formation of music policy. This effectively excluded the DMD from any constructive role in decision-making with regard to the *Cité* and 'forced' Boulez, yet more markedly than in the past, to bypass the DMD and intervene directly with Lang.

Thus, in July 1991, as the *salle modulable* was nearing completion, the question of the hall's 'variable' acoustics, which had been discussed between the *mission musique* and Boulez six years earlier, surfaced once more. When it appeared that Boulez's original demands had been ignored during construction, Boulez insisted that additional tests be carried out by acousticians based at IRCAM, his musical research centre; as a consequence of these tests, Boulez was able to impose substantial modifications to the hall's acoustic system, at a cost of around 20 million francs. Given the need for additional credits to finance this operation, Boulez put his case directly to Lang:

> At that point, I demanded – and I stress, demanded – that if they wanted to have the Ensemble [InterContemporain], I said listen, either you take care of the acoustic problem, or else I'd rather stay at the Cité Universitaire and then good-bye, you can do what you like with your space. ... On that occasion, I used my weight to tip the scales.[45]

Boulez's influence was similarly decisive in the choices that belatedly determined, from 1989 onwards, the *Cité*'s future programming and statutes – an issue that had largely been ignored in the initial phase of the project's elaboration, and that was resolved only very slowly and tortuously. Thus, when an *Association Cité de la Musique* was set up by the Ministry, at the beginning of 1990, to replace the *mission musique* and develop the programming of the future *Cité*, a former administrator of IRCAM, Irène Ajer, was appointed its first, and only, director.[46] Relations between the *association* and the DMD were overshadowed by the increasingly bitter and public conflict between Boulez and Michel Schneider – which culminated in Schneider's resignation in 1991, the ostensible reason being a disagreement with Boulez over the design of the *Cité*'s museum of music. Schneider's successor as *directeur de la Musique*, Thierry Le Roy, member of the *Conseil d'Etat* and Lang's *directeur de cabinet* between 1984 and 1986, was more accommodating towards Boulez. This cleared the way for a more co-operative relationship between Boulez and the DMD, and unblocked the stalemate at the *Cité* which had accompanied Schneider's tenure as *directeur de la Musique*. In November 1991, the preparatory *association* was dissolved and replaced with an *établissement public*, which had the task of overseeing the completion of the concert hall and museum, the split with the *Conservatoire* having been formally consummated. Like the *Association Cité de la Musique*, the *établissement public*, which presided over the *Cité* until the opening of the *salle modulable* in January 1995, bore Boulez's imprint: its director-general, Brigitte Marger, had formerly been administrator of the *Ensemble InterContemporain*. In the words of one DMD official, this reflected not so much Boulez's striving for hegemony as a lack of alternative conceptions on the part of the project's nominal masters:

> Boulez's suggestions for appointing particular individuals ... were followed for want of other candidates. ... Boulez didn't get his own people in so that the project might become what he wanted. No, Boulez simply put his people in because it seemed that there were no candidates, and he said, 'I've got someone at my side who's very good, who could very well do it ...'[47]

CONCLUSION

The brief history of the *Cité de la Musique* presented here has emphasized the central role of the project's professional 'clients' not only in defining the form of the *Cité*, but also in ensuring its completion in an often hostile context. Both the *Cité*'s 'content' and the fact that it survived a series of political and budgetary challenges have to be seen as a function of the influence of its end-users, notably the *Conservatoire* and Pierre Boulez as President of the *Ensemble InterContemporain*. In spite of the discontinuous nature of the *Cité*'s elaboration, the policy-making system that determined the evolution of the project can be usefully characterized using the language of 'policy networks'. It is important, here, to distinguish between different levels of decision-making, and between different elements of the project. Thus, rather than treat the *Cité* simply as a single coherent policy object, we can disaggregate the *Cité* policy arena into a cluster of discrete, if related, policy networks concerned with specific functional components of the project. One such network constituted itself, from 1984 onwards, around the *mission musique* with the *Conservatoire* as a dominant issue. A separate, but connected, network emerged on the basis of another element of the *Cité*, the 'partie Est' and the *salle modulable*, where a pattern of regular and intense interaction developed between the representatives of the core executive, from the *mission musique* to the Minister of Culture, and Pierre Boulez as the principal 'client'. In contrast with the *Conservatoire* network, the 'Boulezian' network was characterized by a high degree of stability and consensus; the network could thus be described as being located towards the 'policy community' end of the continuum suggested by Rhodes and Marsh (1992). Crucially, the existence and strength of the 'core' *Cité* policy community around Pierre Boulez served as a guarantee for the survival of the entire project when faced with external challenges, and provided a rare element of continuity in a policy-making process otherwise marked by drift and uncertainty.

Conversely, the role of the project's nominal political masters was found to diverge markedly from the pattern we might associate with a presidential *grand projet*. The President, far from providing leadership, remained detached from the project; the Minister of Culture, though largely responsible for securing the project's inclusion among the President's *grands projets*, subsequently restricted himself in essence to following the agenda set by Pierre Boulez and other prospective 'clients' of the *Cité*. Of course, while it is argued that extensive professional involvement in policy-making was necessary to the project's success, it is not suggested that it was sufficient: the active support of the Minister of Culture, and the passive acquiescence of the President, provided the political and budgetary context in which a small number of interested professionals could successfully pursue their objectives for the *Cité*. Nor is it implied that the pattern discernible in this particular case is applicable to all, or indeed any, of the other presidential *grands projets*. By definition, policy-making with regard to the *Cité de la Musique* was more susceptible to sectional and often conflictual influences than projects less closely bound up with pre-existing, sectoral cultural policies. However, the contrast between the hierarchical, state-centred form of policy elaboration generally attributed to the *grands travaux*, and the more complex picture emerging from empirical study of this particular *grand projet*, does suggest that the rhetoric and protocol of the *domaine réservé* may in certain cases obscure a more pluralistic policy-making practice. More specifically, the complexity of policy-making patterns observed in the case of the *Cité de la Musique* invites us both to treat with caution generalizing assertions about the 'presidentialization' of cultural policy, and to re-evaluate the role of cultural professionals and policy networks in the decision-making process. To build on these conclusions, and to take the analysis begun in our study of the *Cité de la Musique* further, the chapters that follow will examine patterns of decision-making in public policy towards the performing arts, starting with an area of state intervention which we have already indirectly skirted in this chapter – that is, public support for contemporary music.

4 Contemporary Music Policy

The classification of contemporary music as a distinct sector of cultural policy may at first appear questionable. In France as in most other countries of Western Europe, the 'production' and performance of contemporary 'academic' music are closely integrated with, and dependent on, a traditional musical infrastructure of orchestras, concert venues and teaching institutions. Moreover, the activities generally grouped together under the heading 'contemporary music' are highly differentiated, and range from orchestral performances to electro-acoustic or musicological research with no immediate public application. Nevertheless, it is possible to delineate a policy-making subsystem concerned with contemporary music which, in important respects, is distinct from the traditional classical music sector. First, at a purely administrative level, policy aimed both at the production and performance of new work and at supporting 'musical research' was, for most of the period we are considering, the responsibility of a single department of the *direction de la Musique et de la Danse* (DMD) of the Ministry of Culture. Second, and perhaps more important, most of the institutions concerned with contemporary music, whether performance or research, were established and controlled by a relatively small professional *milieu* of composers and composer-conductors, many of whom shared a similar background of initiation into an 'avant-garde' musical subculture in the 1950s and 1960s. Finally, the contemporary music sector as a whole was characterized by a feature which set it apart from virtually all other sectors of public funding – the extreme paucity of 'lay' audiences, in virtue of which the distribution of the sector's output was often restricted to a small professional community.[1]

Existing accounts of policy towards contemporary music have tended to focus exclusively on this last characteristic, presenting the contemporary music sector as an extreme example of the emergence, in the cultural policy arena, of hermetically isolated, self-perpetuating, professionally dominated clusters of institutions unresponsive to the pressures of audience demand and impervious to political influences. Menger (1983, p. 249), in a definitive sociological study of the post-war contemporary music sector in France, notes that

> a closed production and distribution network has thus been established...: controlled by professionals drawn from the world of musical composition, who nowadays are united at least in considering that any major aesthetic disagreement would be pointless..., this network seems to be in harmony only when protected from all forms of pressure related to public demand, to the point of being in a situation of radical freedom made possible only by the absence of a lay audience.

It could thus be argued that the sector epitomizes better than any other the 'dynamiques endogènes' described by Urfalino (1989) as underpinning the evolution of cultural policies in advanced industrial societies – that is, the apparent absence of any significant societal pressure corresponding to, and explaining, the policy output supplied by government in the policy area in question.

However, the approach taken in this chapter will diverge somewhat from these sociological explanations. Instead of focusing on the sector's undisputed insulation against the laws of supply and demand, we will argue that policy-making with regard to contemporary music has in fact been highly political, in the sense that the distribution of funding has been subject to intense competition between different groups, individuals and institutions. The primary issue to be considered here is the differential treatment afforded the composer and conductor Pierre Boulez – whose role with regard to the *Cité de la Musique* has already been discussed – on the one hand, and the remainder of the contemporary music sector, on the other. Our analysis of the institutional and budgetary imbalance between the 'Boulezian' segment of the sector and other, smaller institutions will then enable us to make more general inferences about the relationship between the structure of the policy-making system and policy outputs in the area of contemporary music.

4.1 BOULEZ: FIRST AMONG EQUALS?

In Chapter 3, we considered the crucial role played by Pierre Boulez in shaping the *Cité de la Musique* project. Boulez's role in this presidential *grand projet* is mirrored and presaged by his prominent, if not hegemonic, position in the determination of policy towards contemporary music. Indeed, Boulez's ability to influence policy-making with regard to the *Cité* could be seen in large part as a spillover from the influence

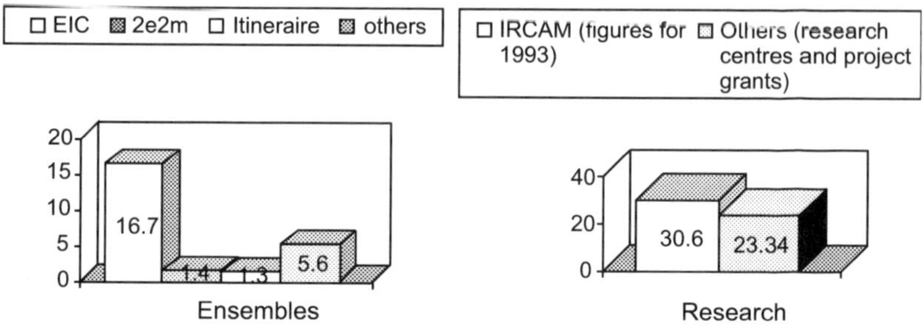

Figure 4.1 Ministry of Culture: allocation of subsidies for contemporary music, 1994 (specialized ensembles and musical research, millions of francs)

Sources: *Diapason*; *Le Monde*.

he acquired over several decades in the area of revenue funding for contemporary music. Since the early 1970s, and irrespective of political changes at the ministerial level, the pre-eminence granted to Boulez and a small number of institutions created by Boulez has been the most immediately striking feature of the state-funded contemporary music sector. Thus, in 1994, even after successive increases in the funding allocated to smaller institutions, the subsidy accorded to IRCAM, the musical research institute founded and headed, until 1992, by Boulez, was greater than the entirety of the subsidies attributed to the remaining institutions in the area of musical research. Similarly, the Ministry's subsidy to Boulez's *Ensemble InterContemporain* (EIC) was twice as great as the budget devoted to the entirety of the remaining ensembles (see Figure 4.1).

IRCAM and the EIC: the Institutionalization of Professional Prestige

Whilst our analysis is focused on policy-making patterns in the period of 1981–93, any attempt to explain the unique position occupied by Pierre Boulez has to consider the historical evolution of Boulez's relationship with the funding system.[2] Throughout the 1950s and 1960s, when Boulez established himself as the figurehead of a post-war musical avant-garde, the composer and his followers remained on the fringes of the state-funded music sector, subsisting mainly through private patronage. However, the musical funding establishment itself increasingly divided into a 'traditionalist' faction hostile to Boulez and a 'modernist' camp who urged that he be given some form of official recognition. In 1966, these tensions came to a head when André Malraux, then Minister for Cultural Affairs, appointed the 'traditionalist' composer Marcel Landowski to the post of *directeur de la Musique* in his Ministry. Reacting to Malraux's decision, Boulez decided to begin a well-publicized 'boycott' of the official funding system, preferring to pursue his conducting career outside France.[3] Malraux's stance against Boulez and the 'modernists' would, a few years later, be reversed – again, by a decision taken at the highest levels of the political executive. Thus, in 1970, President Pompidou invited Boulez to assume responsibility for the musical 'department' of the arts complex then being planned for the Beaubourg site in central Paris, and which would subsequently take Pompidou's name. Boulez agreed – however, his leverage *vis-à-vis* the French President and government had been strengthened as a direct function of his international prestige, and the composer was thus able to impose the terms for his return from 'exile'. Perhaps most important, Boulez insisted that the new institution, the *Institut de Recherche et de Coordination Acoustique/Musique* (IRCAM), have complete autonomy *vis-à-vis* the established musical funding system. Specifically, IRCAM was to be funded directly by the Ministry of Culture as part of its annual grant to the *Centre Pompidou*, rather than receiving its subsidy, like all other nationally funded musical institutions, from the Ministry's *direction de la Musique* (DMD).[4] By contractually tying IRCAM's subsidy to the budget of the *Centre Pompidou*, Boulez obtained a guarantee of long-term budgetary stability – whereas, in the words of one of Boulez's collaborators,

> if IRCAM had found itself confronted directly with the *direction de la Musique*, it would have been necessary to fight all the time. Now one might well say that, IRCAM, at the time, with Pierre Boulez, hardly risked a great deal. But it would have been necessary to spend one's time fighting an intention [on the part of the DMD] to cut funding.[5]

Boulez similarly ensured that IRCAM was generously funded in absolute terms – in its opening year, this meant an endowment of 60 million francs, with a lower but stable annual subsidy in subsequent years. Finally, Boulez asked for undiluted authority over IRCAM's artistic policy. Again, this required insulating IRCAM against the nominal administrative and artistic control exercised over most other institutions by the sectoral bureaucracy. Thus, while the DMD was allowed representation on IRCAM's governing body, none of the officials formally concerned with funding IRCAM were ever able to diminish the autonomy enjoyed by Boulez. As a senior member of Jack Lang's *cabinet* points out,

> in reality, Boulez isn't ultimately under anyone's control. Boulez has three supervisory authorities: there's the *Centre Pompidou*, there's the *direction de la Musique et de la Danse*, and there's the Minister's *cabinet* and the Minister himself. That makes for three different interlocutors. None of the three in fact has any real power over IRCAM.[6]

In parallel with the creation, in 1976, of IRCAM, devoted to musical research and computer-assisted composition, Boulez obtained a commitment on the part of the Ministry to fund a permanent ensemble devoted exclusively to the performance of contemporary music, the *Ensemble InterContemporain* (EIC). Again, as in the case of IRCAM, the EIC was to occupy a privileged place in the Ministry's policy towards contemporary music. While funding for the ensemble, established in 1976, was, unlike IRCAM, devolved to the *direction de la Musique*, the scale and form of the Ministry's support for the EIC contrasted starkly with its policy towards smaller ensembles, some of which had been established several years before the EIC. While the smaller ensembles were initially eligible only for small project grants, the EIC was funded from the outset like a permanent orchestral formation; moreover, the alignment of the EIC musicians' salaries on civil service pay scales effectively precluded significant fluctuations in subsidy. To underline the privileged status of Boulez's ensemble, the EIC was administratively attached – unlike the smaller ensembles – to the section of the DMD responsible for funding large orchestral and operatic institutions.

It was thus clear from the way in which funding for IRCAM and the EIC was set up at the outset that Boulez and the institutions he established would occupy a distinct and privileged place in the contemporary music sector, largely immune to the relative budgetary uncertainty to which ensembles and research centres inside the established musical funding system were prone. The precedent set, and the obligations created, by decisions taken during the Pompidou presidency inevitably constrained the options of policy-makers throughout the late 1970s and 1980s. In order to explore in greater detail the peculiar nature of the relationship between Boulez and policy-makers during the Lang years, we will, first, briefly adumbrate in general terms the structures and procedures by which funding for the EIC and IRCAM was elaborated between 1981 and 1993, and second, consider in greater detail two specific attempts, by two separate *directeurs de la Musique* appointed by Lang, to 'adjust' the perceived imbalance in the structure of the funding system.

4.2 CONTEMPORARY MUSIC POLICY, 1981–93: BOULEZ, IRCAM AND THE EIC

On assuming office in 1981, Jack Lang continued both the policy of maintaining IRCAM and the EIC as prestige institutions outside the remit of conventional music policy, and the tradition of regular personal contact with Boulez at the ministerial level. Indeed, the new Minister's stance was, if anything, more generous than that of his predecessors, given Lang's fundamental emphasis on supporting 'les créateurs' and Boulez's status as France's most acclaimed living musician. According to Boulez, relations with Lang were 'more regular' than with previous ministers, and 'of all ministers [of culture], he was simply the one I saw most often'.[7] However, Boulez did not engage in active 'lobbying' on behalf of IRCAM; according to a source close to Boulez,

> Boulez never asked for a subsidy for IRCAM. If you look closely [at funding], for IRCAM, it's very, very linear. The few benefits that IRCAM obtained, it received directly from the President of the *Centre Georges Pompidou*, who made an effort – because in his case, it is perhaps true that it was important for him to get along well with Boulez. But it never went any further than that.[8]

On a concrete budgetary level, the institutions presided over by Boulez profited handsomely – but not disproportionately – from the doubling of the Ministry's budget in 1981. Thus, IRCAM's subsidy rose from 18.9 million francs in 1981 to 28.3 million francs in 1982,[9] remaining relatively stable in subsequent years (see Figure 4.2).[10]

A different situation obtained with regard to long-term structural decisions affecting IRCAM. Decision-making in this area conformed very clearly to the pattern established in the early 1970s of a close working relationship between Boulez and the Minister of Culture. In a previous chapter, we discussed the role of agreements reached directly by Boulez and Lang with regard to the installation of the EIC as resident ensemble at the *Cité de la Musique*. In the case of IRCAM, important structural decisions were similarly taken in consultation between Boulez and the Minister. This was most obviously the case

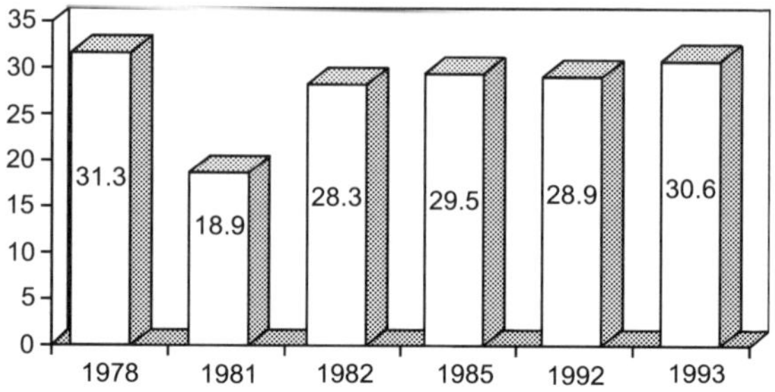

Figure 4.2 Ministry of Culture: annual subsidy to IRCAM, 1978–93 (millions of francs)

Sources: *Ministère de la Culture*; *Le Monde*.

with regard to the construction of a major extension to IRCAM's site next to the *Centre Pompidou*, which was begun in 1987 and the first stage of which was completed in 1989. The case for an expansion of IRCAM – housed, until then, in a subterranean complex of studios, laboratories and offices under a square adjacent to the *Centre Pompidou* – was first put to Jack Lang by Boulez at the beginning of the 1980s. The official decision to construct an annexe above ground, designed by one of the original architects of the *Centre*, was then taken in 1985. The cost of the extension was met in large part by the Ministry of Culture – the first phase, budgeted at around 20 million francs, was funded in its entirety by the Ministry; a third of the budget for the second phase, estimated at around 30 million francs, was to be met by funds raised by IRCAM itself. The role played by Boulez in obtaining the Ministry's support for the extension was, by all accounts, decisive – according to one of Boulez's closest collaborators, 'the construction project would never have come into being' if it had not been for Boulez's intervention.[11] Conversely, the decision to approve the extension was taken quite unambiguously by the Minister of Culture himself, with neither the *cabinet* nor the DMD playing a significant role: as a senior member of Jack Lang's *cabinet* recalls, 'when [Boulez] decided to build his extension – how did it happen? The decision was entirely the Minister's, with the *Centre Pompidou* and the *direction de la Musique* having to obey.'[12] The relative impotence of the 'conventional' agents of intra-sectoral decision-making – the *cabinet* and the administration – is similarly underlined by another senior member of Lang's *cabinet*, who points out that

> you're in a system where you have the impression that when you have a prestige institution like that, you don't have the capacity for counter-expertise, which would enable you to say, 'instead of 30 million, you need 25 million, or 20 million'. It's a question of the balance of power. As in any specialized area like that, you don't really have the means to establish the figures.[13]

The example of IRCAM's extension clearly illustrates the specificity of policy-making with regard to the prestige institutions created by Boulez. In order to cast more light on the nature of the 'Boulezian' sub-sector of contemporary music, and its peculiar relationship with the public funding system, we will now consider the (unsuccessful) attempts of two separate *directeurs de la Musique* in the Lang Ministry, Maurice Fleuret and Michel Schneider, to exercise more direct control over the Ministry's policy towards IRCAM and the EIC.

Maurice Fleuret: the Ideological Challenge

Lang's supportive stance *vis-à-vis* IRCAM, which we discussed above, contrasted with the position of the sectoral bureaucracy – which continued to resent its exclusion from a major sector of the Ministry's music policy. Most important, however, the new *directeur de la Musique* appointed by Lang in 1981, Maurice Fleuret, was fiercely critical of the weight of large institutions in the Ministry's funding policy, and began his term of office by attempting, with varying degrees of success, to adjust the DMD's funding priorities in favour of a more 'democratic' and decentralized policy.[14] In the field of contemporary music, Fleuret unsurprisingly aspired to redress what he perceived as the 'imbalance' between the status of Boulez's prestige institutions and funding for smaller ensembles and research centres; in this, he was publicly supported by a number of leading composers whose own ensembles or studios seemed to suffer from the budgetary priority accorded to the EIC and IRCAM.

However, when Fleuret attempted to translate these concerns into concrete policy changes, he was quickly confronted with the peculiar structure of funding for contemporary music. The latent conflict between Fleuret and Boulez came to a head in the summer of 1982, soon after Fleuret's appointment to the *direction de la Musique*. Specifically, Fleuret gave a press conference at the La Rochelle music festival in June 1982, in which he suggested ending the singularity and dominance of Boulez's institutions, not least by creating several large musical research centres at the expense of IRCAM. Fleuret's challenge provoked an immediate response from Boulez himself. The composer insisted that Fleuret be invited to a meeting between Lang and himself shortly after the La Rochelle press conference; according to a source close to Boulez,

> when Fleuret came to the meeting, Lang doubtless imagined they'd talk about the *Opéra Bastille*, or something like that. And Boulez asked that the meeting begin with another topic, and said, 'I'm going to read you something, Jack, and I'd like to know how you'd react if you were in my place.' And he read the passage from Fleuret's press conference, in front of Fleuret – well, you can imagine the dialogue that followed! And it blocked off any possibility for the *direction de la Musique*, because Fleuret knew only too well that every time, whenever he attacked, there would be a response . . .[15]

Fleuret subsequently tried to exercise at least nominal control over IRCAM and the EIC; however, whenever Fleuret attempted to influence specific internal decisions – such as auditions or the programming of the EIC – he was foiled by Boulez's intransigent refusal to tolerate 'interference' by the DMD. Where disagreement persisted, Boulez was able to intervene directly with Lang, who generally supported the composer – in the words of one of Boulez's closest collaborators,

> [Fleuret's position] never had any concrete consequences for IRCAM, because between 1981 and 1985, this attitude was thwarted entirely by Boulez – who in fact heightened the conflict with the civil servants by refusing to deal with anyone but the Minister.[16]

Paradoxically, then, Lang protected Boulez against the 'reforming' ambitions of the *directeur de la Musique*, whom he had previously appointed not least because of his iconoclastic and radical aspirations for the music funding system. It is interesting to note that Fleuret's concerns were in part shared by some influential members of the *cabinet* – according to a senior civil servant representing the more 'technocratic' elements of the *cabinet*,

> we reckoned that the *direction de la Musique* actually wasn't wrong in thinking that one had to try, in spite of everything, to run a comprehensive policy on contemporary musical creation. This implied that IRCAM, no matter how prestigious and essential it might be, had to

fit into this comprehensive policy. ... Well, at the same time, it's also true that one needed to avoid being on bad terms with Pierre Boulez...[17]

In practice, then, this concern with 'rational' policy-making procedures and administrative control was overruled by the political imperative of avoiding conflict with Boulez.

Another, less obvious, factor also strengthened Boulez's hand *vis-à-vis* Fleuret and the DMD: in the later years of the first Lang Ministry, both Fleuret and Lang were increasingly forced to resort to Boulez's influence to defend the musical *grands projets*. Thus, in the increasingly unfavourable budgetary context that prevailed after 1983, the projects which benefited least from direct presidential patronage – the *Cité de la Musique* and, to a lesser extent, the *Opéra Bastille* – came under strong pressure from the Ministry of Finance; Boulez's active support for the projects, however, was one of the Ministry of Culture's strongest arguments in their favour. As a source close to Boulez points out, 'the further things moved ahead, the more [Fleuret's] position became untenable, because Fleuret himself was calling on Boulez every day to defend the *Opéra Bastille*!... Fleuret, Lang and others were constantly using Boulez as a shield.'[18] Consequently, given that IRCAM and the EIC remained effectively outside his remit, and in view of Boulez's strategic importance as an influential supporter of the musical *grands projets*, Fleuret could do little more than attempt to redress the imbalance between Boulez and the rest of the sector by increasing, as much as possible, the subsidies granted to the smaller ensembles and research centres.

Michel Schneider: the Bureaucratic Challenge

As we have seen, Maurice Fleuret's tentative attempt to attenuate the perceived 'Boulezian' hegemony in the contemporary music sector was 'defused' before a more open conflict between Boulez and the *direction de la Musique* could develop. The resulting 'truce' between Boulez and Fleuret contrasts sharply with the outcome of a more violent, and public, conflict between Boulez and Michel Schneider, the *directeur de la Musique* appointed by Lang on his return to the Ministry of Culture in 1988. Schneider's disagreements with Boulez were not focused directly on funding for IRCAM and the EIC, as the conflict with Fleuret had been. However, as they throw further light on the nature of Boulez's relationship with policy-makers, and notably the

Minister of Culture, it will be helpful briefly to sketch the evolution of the Schneider 'affair'.

Schneider's conflict with Boulez was primarily focused on Boulez's role with regard to the *grands travaux*. Boulez accused Schneider of indifference to the problems encountered in the elaboration of the final programme for the *Cité de la Musique*, while Schneider resented the pre-eminent role played in the decision-making process by an individual who was, officially, little more than one of several eventual end-users of the *Cité*.[19] More generally, Schneider insisted, in his dealings with both Boulez and other recipients of public funding, on scrupulous adherence to what he saw as the necessary separation of responsibilities between the state and its 'clients'. In other words, Schneider perceived his role as that of a guardian of the impartiality and independence of the state, and more specifically of the administration as guarantor of the 'general interest'.[20] Boulez, on the other hand, clearly accorded the administration a subordinate role, purely 'supportive' of the *créateur*.[21] The conflict crystallized from 1989 onwards around an exchange of letters between Schneider and Boulez in which they frankly stated their positions. According to a source close to Boulez,

> things became very conflictual after that, because in addition Boulez's response had been in much more violent terms. And Boulez said all this in front of everyone, because he took the letter from Schneider – who had wanted all this to remain confidential – and then he took his own letter, and he sent the lot to Lang.[22]

Indeed, Boulez explicitly entreated Lang to divest himself of his *directeur de la Musique*. In the composer's own words,

> in the case of a *directeur de la Musique* like Schneider, [Lang] should have acted much more quickly. I told him after one year, 'Be careful, this isn't going to work', and I gave him some concrete examples of decisions, etc. And Lang was more or less conciliatory, [saying] 'Schneider isn't really a very important person in the hierarchy' – but he wasn't willing [to act]. And in the end, he got all the trouble, much more than he would have got if he'd discreetly fired him at the beginning. Because at that point, a clash inevitably occurred.[23]

The impact of the conflict between Boulez and Schneider, which ended only with Schneider's resignation, was, as we saw in a previous chapter, felt most tangibly in the elaboration of the *Cité de la Musique*. In the sector of contemporary music as such, the effects of Schneider's tenure were less perceptible. Funding for IRCAM and the EIC was unaffected. In any case, even if Schneider had wanted to adjust the existing distribution of subsidies, he lacked the resources to impose any change in policy – in the words of one DMD official, 'he would certainly have liked to, but he didn't have the political means to do it.'[24]

We may note that there are strong parallels between the unsuccessful attempts of Maurice Fleuret and Michel Schneider to impose a change of policy towards Boulez's status in the music funding system. In Fleuret's case, a specific 'ideological' commitment to adjusting the perceived hegemony of IRCAM and the EIC was rapidly dropped as it became apparent that the *directeur de la Musique* had no leverage over decisions relating to the two 'Boulezian' institutions. Michel Schneider, motivated less by any substantive policy aim than by a general concern to enforce the primacy and autonomy of the administration *vis-à-vis* its 'clients', similarly failed when he challenged the autonomy enjoyed by Boulez. While the conflict between Schneider and Boulez focused ostensibly on the *Cité de la Musique*, the vigorous action taken by Boulez, Schneider's failure, and the relative passivity of Jack Lang in resolving the controversy all offer a clear illustration of the underlying distribution of influence that conditioned the Ministry's policy towards the 'Boulezian' segment of the contemporary music sector. The specificity of the policy-making patterns prevailing in this area was further borne out by the evolution, over the same period, of the Ministry's policy towards other, smaller ensembles and musical research centres, to which we now turn.

4.3 CONTEMPORARY MUSIC POLICY, 1981–93: *'LES AUTRES'*

IRCAM and the EIC were by no means the only institutions of their kind when they were created on Pierre Boulez's 'return' to France in the early 1970s. A number of electro-acoustic studios were established from the late 1960s onwards, partly under the auspices of the ORTF broadcasting network, partly on the private initiative of individual composers. While often labelled 'research' institutes, many of these studios were in fact tailored to the compositional needs of their founders. At the same time, a number of chamber ensembles devoted entirely to contemporary music were established; the best-known, the Paris-based 2e2m and *L'Itinéraire*, were created in 1972 and 1973, respectively. As in the case of many of the electro-acoustic studios, most of the ensembles were closely tied to particular composer-conductors. Funding for both the ensembles and the research centres had been initiated during Marcel Landowski's tenure as *directeur de la Musique* in the early 1970s; however, until 1981, subsidies remained at a symbolic level, and the volume of funds allocated to contemporary music actually diminished during the Giscard years. In 1980, the musical research centres supported by the *direction de la Musique* received subsidies totalling no more than 2.65 million francs, while the budget allocated to specialized ensembles and contemporary music festivals amounted to 0.59 million francs.[25] The impact of the near-doubling of the DMD's budget in 1981–2, coupled with Maurice Fleuret's accession to the post of *directeur de la Musique*, was immediate and dramatic: funding for musical research was multiplied by five, while subsidies for ensembles and festivals were increased sixfold. By 1985, the budget for musical research had risen to 19.01 million francs, while the ensembles received a total of 6.81 million francs; overall, the Ministry's subsidies to contemporary music, excluding IRCAM and the EIC, rose from 6.47 million francs in 1980 to 40.04 million francs in 1985.[26] The increase in funding was accompanied by an expansion in the number of institutions supported, particularly in the domain of musical research, where the DMD actively encouraged the creation of around 20 new research centres.

The decision to initiate a massive expansion in funding for contemporary music can be attributed almost entirely to Maurice Fleuret. As we saw in discussing the conflict between Fleuret and Boulez, the new *directeur de la Musique* was motivated in part by a desire to redress the funding imbalance between Boulez and other composers and institutions. However, Fleuret, who had developed close ties with the contemporary music *milieu* long before 1981,[27] also had a personal interest in the sector, no doubt more so than in any other funding area in his remit. Thus, numerous individual funding decisions relating to specific composers or institutions were taken by the *directeur de la Musique* himself. By contrast, both the Minister and his *cabinet* remained detached from policy-making. Jack Lang intervened on the very rare occasions when particular funding decisions intersected with explicitly political concerns; thus, the creation, in 1983, of Strasbourg's 'Musica' contemporary music festival, in a region controlled by the Right, required support at the ministerial level to secure some degree of local government funding. Yet given the small scale and relative obscurity of most of the institutions concerned, such interventions were exceptional – in contrast with central-local government relations in more sensitive areas, such as the funding of regional orchestras or, to take an example from another sector considered in this book, regional theatre companies.

In any case, Lang's objectives in the area of music policy were only partially congruent with Fleuret's preoccupations. While Fleuret had long been, in the words of one DMD official, an 'unconditional, very strong, very active and almost militant supporter of the young contemporary music scene', Lang was seen by those inside the sector to be more concerned with the eventual political benefits of his music policy: 'It's not necessarily a sector that Lang pampered, because it didn't bring him the kind of publicity... that he could get much more easily with the singers, the rock stars, etc.'[28] Consequently, Fleuret's status in the Ministry ultimately depended on the implicit understanding that, in addition to the policies to which Fleuret was personally most strongly committed, the DMD would pursue a visible and politically effective policy towards jazz, rock and *variétés*. In the area of contemporary music, however, the DMD was effectively given *carte blanche*. Under Fleuret, funding decisions in many instances reflected the personal priorities of the *directeur de la Musique*; under subsequent *directeurs de la Musique*, whose personal interest in contemporary music was less pronounced, responsibility for the sector was effectively devolved to lower-level civil servants.

The account set out above thus points to a strong role for the sectoral bureaucracy in the definition both of the overall level of support and of the distribution of funds between individual institutions. Conversely, we established that the Minister and his *cabinet* played virtually no role in the definition of policy towards contemporary music ensembles and research centres. In short, the distribution of roles between the political executive and the administration was the reverse of that observed in the case of the 'Boulezian' sub-sector. However, our analysis of the policy style prevailing in this sector will remain incomplete if we do not take account of the role played by the profession itself in the elaboration of policy. As we shall see, there is a marked contrast here, too, with the role played by Pierre Boulez.

Professional Inputs into Contemporary Music Policy

As suggested earlier, most of the specialized ensembles and electroacoustic research centres supported by the Ministry had originally been established by, and continued to be headed by, individual composers.[29] Indeed, given the absence of either substantial audiences or a significant capacity for attracting media attention, the personal prestige of their artistic directors was the only resource the institutions concerned could draw on in their dealings with the Ministry. During Fleuret's tenure at the DMD, the sector benefited from the personal 'protection' of the *directeur de la Musique*. Under Fleuret's successors, however, whilst a very small number of composers with a wider public reputation retained a certain 'immunity' to dramatic fluctuations in their subsidy, the weakness of the majority of professionals in the sector became more evident. According to one DMD official, 'nobody cares about this sector. The clout of the composers is too feeble. They're isolated individuals who tried at one point to unite and form a kind of movement, but it's all too weak to have an impact.'[30] In fact, the sector was characterized, whether in the case of the ensembles or with regard to the research centres, by the absence, in the period considered, of any serious attempt to organize a collective strategy *vis-à-vis* the Ministry. As the administrator of one ensemble points out, 'each ensemble tries more or less to defend its own policy. ... It's a very fragmented *milieu*.'[31] The futility of rare attempts to 'federate' the *milieu* only served to underline its intrinsic fragmentation. Thus, under Lang's conservative successor, Jacques Toubon, a 'Comité de co-ordination Musique-Danse', inspired by the example of the SYNDEAC, the very effective theatre directors' union, was created to protest against a series of budgetary cuts proposed for the contemporary music sector in the summer of 1994. However, the campaign instigated by the 'Comité' never advanced beyond a circular to the administrators of the ensembles and research centres concerned – which, characteristically, was ignored by most of its recipients. In the words of one of the latter, 'it was a pointless exercise ... it wasn't very serious. Perhaps it was meant yet again to save particular individuals.'[32]

The fragmentation of the profession – which sets the world of contemporary music apart from the other policy sectors we are considering – inevitably made individual institutions more vulnerable to budgetary fluctuations and weakened the capacity of the sector as a whole to obtain budgetary advances. In the case of the ensembles, this was reflected most obviously in the stagnation, and in some cases decline, of annual subsidies after the initial expansionary phase under Maurice Fleuret – thus, the annual subsidy of 2e2m, one of the two ensembles supported by the Ministry on a permanent, revenue-funded basis, declined by 28 per cent in real terms between 1982 and 1991.[33] The fragility of individual institutions, and their relative weakness *vis-à-vis* the Ministry, were even more apparent, however, in the area of musical research.

Musical Research Policy: Bureaucratic Voluntarism and Professional Vulnerability

In 1982, as we noted earlier, Maurice Fleuret initiated a dramatic expansion in the number of electro-acoustic research centres, which absorbed much of the sudden budgetary increase granted to the contemporary music sector in the 1982 budget. On Jack Lang's return to the Ministry in 1988, however, the DMD initiated a significant shift in its policy towards musical research. The rationale underlying Fleuret's initial decision had, by now, come to appear questionable. Indeed, the notion that 'musical research' as such constituted a coherent policy object had proved difficult to maintain in practice. Some of the programmes and institutions funded under the general heading of 'research' were concerned with 'traditional' musicology. Most of the 'research' budget, however, was marked for electro-acoustic 'research', a category which, moreover, had benefited most from the 1981–2 expansion. Yet it was in the area of electro-acoustics that the term 'research' was most problematic: many of the composers who had been granted funding to establish or consolidate electro-acoustic studios were in fact using these facilities to further their own compositional activities. In 1989, in a report commissioned by the DMD, Pierre-Michel Menger thus underlined the difficulties of distinguishing '*recherche*' and '*création*', and suggested that the notion of musical research, as used until then by the DMD, might correspond to no more than 'politico-administrative fiction' (Menger, 1989, p. 163). Moreover, Menger observed that the criteria employed by the DMD in determining the recipients of research funding were extremely loose – once a studio had entered the DMD's 'circuit', permanent funding was effectively assured.

In 1989, partly in response to Menger's report, the DMD implemented a number of reforms which effectively restructured the research sector. At the administrative level, the *division* which had been responsible both for research and for 'création' – that is, for the specialized ensembles, festivals and public commissions – was split in two, with separate *divisions* responsible for research and 'création', respectively. At the same time, an advisory commission of external experts was created with the aim of accentuating the 'scientific' dimension of the Ministry's policy on musical research. Perhaps most important, the DMD began progressively to reduce the number of research centres it subsidized. Under Fleuret's tenure as *directeur de la Musique*, the sector had expanded from the six centres initially supported to 26 at the end of the first Lang Ministry; by the end of the second Lang Ministry, ironically, the DMD had terminated its funding to all but six centres. The reform was tantamount to an admission that the expansionary policy pursued in the early 1980s had been based on false premises: in the words of a senior DMD official, 'even if the people working in the sector don't really like being told this, it's true that one used to engage in musical research to get a bit more money. You didn't produce a thesis, you didn't produce any written work, you didn't have any obligation to answer for what you were doing.'[34] Implicitly, the new policy also recognized the benefits of large, adequately funded institutions modelled, ultimately, on IRCAM – in complete contradiction with the principles which had driven Fleuret's policy. As a source close to Pierre Boulez recalls,

> from 1989–90, you had a situation where even the 'left-wing' Ministry pushed [IRCAM] into being the only important centre. We even had to calm them down, and we found ourselves ... in the situation of having to defend, on several occasions, the existence of small research centres belonging to particular associations – because the Ministry's idea was to abolish, abolish, abolish![35]

Remarkably, policy towards musical research thus presents a pattern of relations between the administration and a professional *milieu* in which the former was clearly able to define and implement a set of policy objectives – in this case, the 'rationalization' of the research sector and the concentration of resources on a small number of viable institutions – even where this entailed terminating subsidy to a large number of 'clients'. Of course, the significance of this example should not be exaggerated. The volume of funds involved pales beside the subsidies accorded to institutions in the traditional performing arts sectors: in 1991, the budget of the entire *division* responsible for research and audiovisual policy amounted to no more than 22 million francs, with most individual subsidies not exceeding 200,000 francs. Nonetheless, the process described clearly stands out as an example of 'hierarchical' policy-making in a policy area which, overall, was characterized by more horizontal modes of interaction between policy-makers and their 'clients'. It thus underlines the relative weakness of professional interests in the wider contemporary music sector, and enables us to draw more general inferences about the determinants of policy in this area, both in the 'Boulezian' segment of contemporary music and with regard to the smaller institutions supported by the *direction de la Musique*.

CONCLUSION

We noted, at the beginning of this chapter, that the imbalance between the two institutions founded and patronized by Pierre Boulez and the remainder of the sector was the most immediately striking feature of the allocation of public funding for contemporary music. In considering the policy-making processes that lay behind these distributive patterns, we found that the two segments of the sector – a 'Boulezian' complex and the less privileged institutions – corresponded to two discrete and perfectly isolated policy-making sub-systems. The first, concerned with IRCAM and the EIC, was treated exclusively at the level of the Minister of Culture and was effectively inaccessible to the sectoral bureaucracy nominally responsible for contemporary music. A satisfactory progression of the two institutions' annual subsidies was assured by statutory devices which protected them from the budgetary fluctuations to which funding for contemporary music had traditionally been subject. Important structural questions affecting either of the two institutions – such as IRCAM's extension or the EIC's move to the *Cité de la Musique* – were submitted directly to Jack Lang by Pierre Boulez and decided by the Minister without either the *cabinet* or the bureaucracy playing a significant part. It should be underlined, however, that the policy agenda was set exclusively by Boulez; the Minister's role was essentially one of following this agenda and approving the disbursement of funds where considered most appropriate by Boulez. The insulation of IRCAM and the EIC against 'external' challenges was underlined, and reinforced, by the failed attempts of two of Lang's *directeurs de la Musique*, Maurice Fleuret and Michel Schneider, to tackle what they perceived as Boulez's hegemonic status in the contemporary music funding system.

A mirror-image of this pattern obtained, on the other hand, with regard to policy towards the smaller ensembles and research centres. Here, we found that the elaboration of policy was entirely in the hands of the sectoral bureaucracy, and the sudden expansion of the sector in 1981–2 was almost wholly attributable to Lang's first *directeur de la Musique*, Maurice Fleuret, who conceived the DMD's new policy in part as an indirect means of challenging the pre-eminence of the

'Boulezian' segment of the sector. The Minister and his *cabinet*, by contrast, were absent from policy-making in this area. Finally, we noted that the ensembles and research centres supported by the DMD were, in general, in a position of weakness and dependence *vis-à-vis* their supervisory authority. Given the inability of most of the institutions and individuals concerned to draw on significant lay audiences or on the informal support of a wider politico-administrative or economic elite, the allocation of satisfactory subsidies depended largely on the personal backing of the *directeur de la Musique*, both for the sector as a whole and for individual ensembles and centres. During Fleuret's tenure at the DMD, contemporary music – even beyond the confines of Boulez's 'prestige' institutions – was clearly a policy priority. Under subsequent *directeurs*, this was less obviously the case, and the consequent budgetary fluctuations to which individual institutions were subjected served to underline the relative weakness of the profession. In short, the contemporary music sector comprised two discrete 'networks' with very different characteristics; the nature of the relationship between the two networks will emerge more clearly if we apply some of the conceptual tools developed in the policy network literature.

Policy Communities and Issue Networks: 'Insiders' and 'Outsiders' in the Contemporary Music Sector

In applying the network model to the interaction of public and private sector actors involved in the elaboration of policy towards contemporary music, it is important, first of all, to distinguish between the internal structure of the two discrete segments which make up the sector, and the sector as a whole. The patterns observed in each of the two segments were relatively straightforward. The relationship between the Ministry and the various smaller institutions subsidized by the DMD is best captured by the issue network concept, as defined by Rhodes and Marsh (1992, p. 187): the network comprised a wide range of interests, and was based on fluctuating degrees of contact and access; the underlying relationship between different members of the network was consultative. Moreover, the distribution of power within the network was far from equal – in this case, the representatives of the Ministry were clearly dominant and thus able to impose outcomes on their 'clients'. Moreover, even if distributive issues were rarely aired openly inside the network, the funding system can be described as a zero-sum game, with acute competition between different 'clients' for scarce resources. On the other hand, the relationship between Boulez and relevant public officials – in this case, the Minister of Culture himself – can be said to conform loosely to the policy community concept as defined by Rhodes and Marsh (1992, p. 187), with frequent high-quality interaction on all important issues, and a consensus on policy objectives among different members of the network. There are, of course, limitations to the acuity of the network approach in this context. It could be argued that to construe the interaction between two powerful individuals – Boulez and the Minister – as a 'network' is, by definition, absurd as it abuses a concept designed for fundamentally different kinds of structure. Nevertheless, the policy community concept helps to isolate the most essential structural elements of the policy-making system we are concerned with here. Specifically, the policy community concept may help us make sense of the underlying rationale of the relationship between Boulez and successive Ministers of Culture, by pointing to the 'exchange' of resources as a crucial prerequisite of a stable network. The 'exchange' involved is not, of course, quantifiable. While certain 'prestige' investments in the cultural domain – such as, in the French case, the *Opéra Bastille* – may offer tangible returns to their patrons, this is not obviously the case in the area of contemporary music. Rather, the benefits accruing to the political executive are likely to be of an intangible nature – in the case of Boulez, a 'legitimating' effect on perceptions of the Ministry of Culture, and a 'guarantee' against the loss of 'legitimacy' which might ensue if Boulez decided to withdraw yet again from the official music funding system. By contrast, we might note the explanatory limitations of simple hierarchical concepts like 'ministerial government', which fail to capture the centrality of 'exchange' and interdependence to the stable relationship between Boulez and his ministerial patrons.

The structure of the overall contemporary music sector is less straightforward. Indeed, it could be argued that the 'Boulezian' segment and the smaller institutions constitute two entirely distinct and unrelated sectors – a conception to which Boulez, for example, would subscribe. Nonetheless, besides the obvious substantive similarities between the two segments, the very presence of conflict between Boulez and the sectoral bureaucracy – which, as we saw, recurred frequently throughout the period considered – and the persistent ambition of both the DMD and members of the ministerial *cabinet* to impose a more unified policy towards contemporary music enables us to conceive of the two segments as constituting a single sector. The relationship between the different components of the 'network' concerned with policy in this sector has, of course, been the central theme of this chapter. Again, the issue network concept offers the most appropriate way of picturing its underlying structure – that is, one of fundamental inequality of access to political support and resources, and one accommodating several discrete and dissimilar sub-systems. As we shall note in the chapters which follow, this pattern was not restricted to the contemporary music sector.

5 Theatre Policy

Of all the policy sectors considered in this book, the subsidized theatre could be most readily expected to exemplify a specifically French mode of relations between the state and cultural professionals. The historical roots of state intervention go back further in the case of theatre than in any other sector. The *Comédie Française*, the most prestigious theatrical institution in France, was founded in 1680, under the patronage of Louis XIV, and subsequently the role of central government in funding and regulating the theatre has expanded steadily. In the twentieth century, the consolidation of a national cultural policy, notably since 1959, has itself been centred on the development of a large subsidized theatre sector. Both the notion of cultural 'decentralization', in the sense of developing a provincial network of cultural venues equivalent to the great Parisian institutions, and the concept of an active policy of 'democratization', or *action culturelle*, originated in, and were first applied in, the theatre sector. In budgetary terms, too, the theatre has weighed heavily among the priorities of cultural policy-makers: in the period under consideration, funds accorded the theatre sector accounted for, on average, around 13 per cent of the overall budget of the Ministry of Culture.

Yet perhaps the most important argument for devoting a chapter to the theatre policy pursued under the Lang Ministry stems from the formal organization of the subsidized theatre in France. As a brief presentation of the main institutional traits of the sector will emphasize, relations between the Ministry of Culture and theatrical institutions appear, formally, to conform closely to a statist, hierarchical conception of the cultural policy process. The Ministry of Culture could thus be seen to play an active, indeed dominant, role, not only in the provision of venues and funding of theatrical production, but also in shaping the artistic orientation of the subsidized theatre, notably by means of far-reaching prerogatives in the appointment of senior artistic and administrative personnel. In fact, as our analysis of theatre policy-making in the 1981–93 period will indicate, the theatre sector exemplified a very different style of policy-making. Far from being in a position to impose a coherent set of policy objectives on the sector, or even to bear significantly on individual artistic choices, policy-makers were severely constrained by the demands of a highly organized and influential profession, and by a strong informal tradition of consensual policy-making based on close collaboration between senior officials and leading professionals.

Our analysis will focus on three specific aspects of theatre policy: the elaboration of the overall budget allocated to the theatre sector by the Ministry, and the distribution of these funds to different 'clients'; the Ministry's relations with the *compagnies dramatiques*, or independent companies; and the process by which the directors of subsidized institutions were appointed. Our account of each of these will underline the fundamental contrast between the institutional form of public intervention on the one hand, and the reality of a finely balanced collaborative relationship between the Ministry of Culture and its 'clients', on the other. We begin, however, with an overview of the overall structure and organization of the theatre sector, with particular reference to changes which occurred during the 'Lang years'.

5.1 PUBLIC FUNDING OF THE THEATRE, 1981–93

As we noted earlier, systematic public arts funding began earlier in the theatre than in any other sector of French cultural policy. The long tradition of state intervention inevitably left its marks on the institutional structure of the sector, and continued to condition the choices of policy-makers throughout the 'Lang years'. Two formal features of the subsidized theatre exemplify most strongly the statist tradition in theatre funding: the hierarchical structure of the sector, and the extensive administrative and artistic control over individual institutions exercised, at least formally, by the theatre directorate, the *direction du Théâtre et des Spectacles* (DTS), of the Ministry of Culture.

At an organizational level, perhaps the most striking feature of the French theatre sector is its rigorous stratification. Throughout France, theatrical institutions are divided into four distinct institutional categories corresponding, in theory at least, to a ranking order of artistic 'importance' and thus eligibility for subsidy. At the top of the pyramid, five *théâtres nationaux*,[1] all but one of which are based in Paris, are funded exclusively by central government, have the legal status of *établissements publics*, and are placed directly under the Ministry's authority – indeed, the Administrator of the *Comédie Française*, the oldest of the *théâtres nationaux*, is appointed by a prime ministerial decree countersigned by the President of the Republic. Below the *théâtres nationaux*, a second 'tier' is composed of a number of mainly provincial institutions known as *centres dramatiques nationaux* (CDN).

Although the CDN receive some funding from their respective local authorities, they are tied contractually to the Ministry of Culture, and the Ministry has a decisive role in the appointment of the CDNs' directors.[2] In 1981, there were 23 *centres dramatiques nationaux*; by 1990, the CDN sub-sector had grown to 42 theatres. A third tier, which only became formally attached to the theatre sector in 1986, consists of the *maisons de la culture* and *centres d'action culturelle*. Between 1981 and 1986, these multidisciplinary institutions created in the 1960s and 1970s on the initiative of André Malraux, and often referred to as *établissements d'action culturelle*, had been attached to the Ministry's cultural development directorate, the *direction du Développement culturel* (DDC); we will discuss the institutional conflicts surrounding this segment of the theatre sector in greater detail in a separate chapter on *développement culturel*. The last 'tier' in the institutional hierarchy comprises several hundred *compagnies dramatiques*, or independent companies, generally lacking a permanent institutional base and dependent for their productions on co-operation with a CDN or *maison de la culture*. Within this category, there is a further subdivision between companies funded on a temporary basis, and subject to regular evaluation by an advisory commission, and a more privileged class of companies, designated by the term *hors commission*, which are guaranteed subsidy on a permanent basis.

The institutional structure of the subsidized theatre thus takes the shape of a pyramid; conversely, given the funding differentials between the four different types of institution, notably the heavy bias towards the five *théâtres nationaux*, the funding structure of the theatre sector has traditionally resembled an inverted pyramid. After the appointment of Jack Lang as Minister of Culture, the budgetary dominance of the *théâtres nationaux* was diminished significantly: the new Minister thus decided almost to triple funding for the independent companies, entailing a more even distribution of budgetary allocations between the different categories of institution (see Figure 5.1).

As in other sectors, the new emphasis on supporting previously neglected areas of cultural production was counterbalanced by the consolidation of larger, established institutions. The multiplication of subsidies to the independent companies was thus funded not by a redistribution of resources within the sector, but by a large increase in the overall theatre budget, which benefited all segments of the subsidized theatre. Subsidies for the five *théâtres nationaux* were increased by almost 40 per cent, while the funds available to the CDN were doubled. In subsequent budget years, by contrast, the scope for

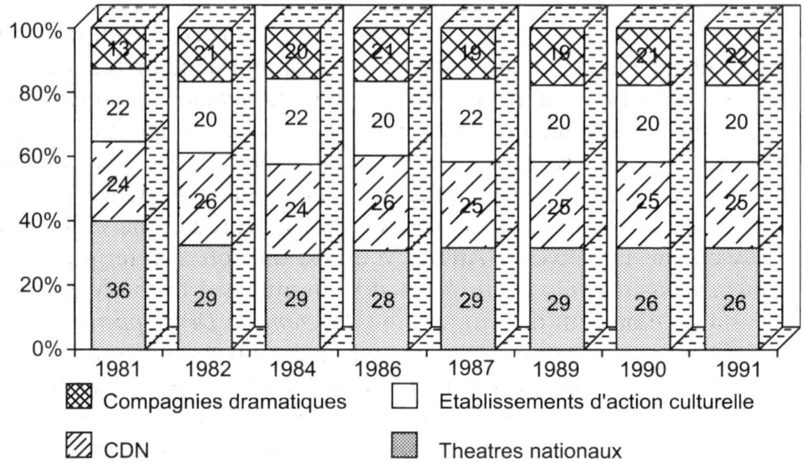

Figure 5.1 *Direction du Théâtre et des Spectacles*: breakdown of budget by category of institution, 1981–91 (% of total, *fonctionnement* only)

Source: *Ministère de la Culture.*

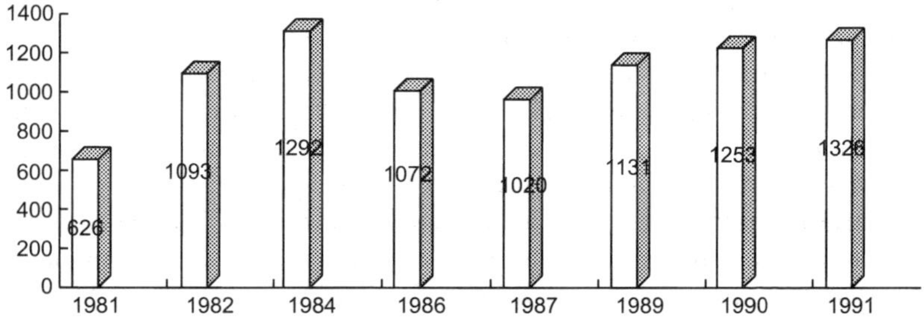

Figure 5.2 Budgetary allocations to the theatre, 1981–91 (*dépenses ordinaires* and *autorisations de programme*, including *établissements d'action culturelle*, millions of francs, indexed to 1991)

Source: *Ministère de la Culture*.

expansion had all but disappeared. Only the 1989 budget would, after a significant reduction of the Ministry's funds during the *cohabitation* years, once again offer the Ministry a small amount of financial leeway, with an increase of around 11 per cent in the overall theatre budget (see Figure 5.2). However, the expansion of 1989 has to be set against a decline, in real terms, of the theatre budget by more than 10 per cent between 1984 and 1988 – overall, there was thus very little scope for major budgetary adjustments after 1981, and the increase obtained in 1989 was spread more or less evenly among the Ministry's client institutions.

While the 1982 budgetary expansion allowed for a significant shift in the Ministry's funding priorities, the remainder of Lang's tenure at the Ministry of Culture was marked by an exceptional degree of stability in the distribution of funding between different sub-sectors and institutions in the theatre sector. The only structurally significant change to occur after 1981–2 concerned the formal transfer of the *maisons de la culture* and *centres d'action culturelle*, initially controlled by the *direction du Développement culturel*, to the theatre sector – a decision taken in 1986 by François Léotard, Minister of Culture during the first *cohabitation*, but subsequently confirmed by Jack Lang on his return to the Ministry of Culture. At the level of individual institutions, the theatre policy pursued by Lang was somewhat more 'activist': thus, a number of new *centres dramatiques nationaux* were created, usually by elevating an existing independent company to the status of CDN. Moreover, Lang made extensive use of his formal power over artistic appointments to the *théâtres nationaux* and the CDN. Between 1981 and 1988, the directors of more than two-thirds of all CDN were replaced; we shall return to, and reassess, the significance of these measures in a subsequent section of this chapter.

Our brief presentation of theatre policy between 1981 and 1993 should have served, above all, to underline the contrast between the ostensibly vast prerogatives of the Ministry of Culture, on the one hand, and the relative stability of funding patterns after 1981, on the other. In fact, as we suggested earlier, the continuity of policy outputs coincided with a pattern of close collaboration between public policy-makers and a highly organized and influential theatre profession. A closer examination of relations between the Ministry of Culture and its 'clients' in three specific areas of theatre policy will enable us to clarify further the link between the form of policy-making and the shape of policy outcomes. We begin by considering the factors that influenced the elaboration of the Ministry's theatre budget.

5.2 THE THEATRE BUDGET, 1981–93

As noted above, both the overall volume of funds and the distribution of resources between different categories of institution remained remarkably stable throughout the 'Lang years'. In order to throw light on the exceptional stability of budgetary policy-making in the theatre sector, which is underscored by any comparison with other cultural policy sectors, we will proceed in two stages: first, we will consider the initial 'round' of allocations in 1981–2, which inevitably conditioned subsequent budgetary choices; and second, we will assess the specific influences which constrained the Ministry's ability to impose significant policy changes after 1982.

The 1982 Budget and the Theatre Sector

Like its counterparts in other sectors, the Ministry's *direction du Théâtre et des Spectacles* was confronted, in the summer of 1981, with the necessity to distribute a substantial increase in budgetary resources in an exceptionally short period of time. The overall volume of funds allocated to the theatre sector after Lang had obtained a near-doubling of his Ministry's budget represented an increase of 80 per cent over the previous year's budget. The entirety of the increase had to be allocated by the end of August 1981. In this context, it might have been expected that the new Minister would use his prerogatives and play an active role in the internal allocation process – not least in view of Jack Lang's background as founder of the Nancy theatre festival and, later, director of the *Théâtre National de Chaillot*. In fact, ministerial involvement was limited – in the words of Robert Abirached, appointed head of the *direction du Théâtre* after Jack Lang's arrival at the Ministry of Culture,[3]

> Lang virtually gave me *carte blanche*, and I had the loyalty and the intelligence to keep him informed of any problems and any questions which might arise.... But one mustn't at all imagine that Lang would take his pen and allocate funding himself.[4]

Abirached's distributive policy, on the other hand, consisted essentially in spreading the funds obtained by Lang evenly among different sub-sectors and institutions, consolidating the status of the *théâtres nationaux* and the CDN while satisfying the demand for financial recognition which had been articulated with increasing urgency by the independent companies. To a certain extent, Abirached's choices were predetermined by funding commitments made by the Ministry before 1981, which had remained unfulfilled under successive Giscardian Ministers owing to the shortage of funds. Thus, in the case of the independent companies sub-sector, a 'backlog' of almost 70 companies recommended for upgrading to permanent or *hors commission* funding status had accumulated in the late 1970s; the first important decision on the part of the new Minister and his *directeur du Théâtre* thus consisted in rapidly expanding the number of *hors commission* companies, concomitantly raising the proportion of funds allocated to the independent companies within the overall theatre budget. Similarly, the substantial increases in revenue funding granted to many CDN and *théâtres nationaux* were largely accounted for by the need to compensate for the diminution in subsidy which many institutions had sustained during the Giscard presidency. According to Abirached, the undiscriminating budgetary magnanimity of 1981–2 owed much to an 'extraordinary pressure' emanating from the theatre profession:

> The profession as a whole had been kept on a very short leash, and felt – very reasonably, as it happens – that for two ot three years the theatre had been particularly neglected and funding had declined … . So expectations were running extraordinarily high. And the cultural world had almost without exception voted for Mitterrand. Which means there was an expectation – it was impossible to do nothing.[5]

Yet while most distributive decisions were left, in this crucial period, to the *directeur du Théâtre*, and while the options of the *directeur du Théâtre* were circumscribed by the imperative of compensating for budgetary restrictions imposed under previous Ministers, Lang did occasionally intervene on behalf of specific institutions. In many cases, Lang's wide-ranging network of personal acquaintances in the theatre world played an important role in facilitating access to the *direction du Théâtre*, if not the actual disbursement of subsidies: as one director of an independent company recounts,

> I saw him [Lang] in October 1981 …, there was a little reception …, and he was Minister, and he said to me, 'Right, what can I do for you? What would you be interested in?' And I told him, 'What I'd be interested in is having some money so I can work with my company.' … And so he says, 'OK, talk to Robert [Abirached] about it.' And it's true, he did this with lots of people …[6]

This pattern of decision-making, whereby budgetary allocations were generally considered a matter for the *directeur du Théâtre*, while the Minister reserved the right to intervene on behalf of particular institutions or individuals, persisted in subsequent years. According to a *cabinet* member responsible for the theatre sector under the second

Lang Ministry, the Minister generally made only marginal alterations to the budgetary proposals emanating from the administration: 'There were some minor changes in the final budgetary discussions with regard to what the *direction du Théâtre* had proposed.'[7] Similarly, according to a senior official at the *direction du Théâtre*, the Minister was

> very involved with regard to the overall volume of the budget, but not really with regard to the allocation of funds. He left a lot of scope, apart from some minor and very specific issues – because he had met such and such a person who had told him about a very particular problem, for which he would then write a note ...[8]

In a few, very exceptional cases, even the President of the Republic intervened on behalf of particular artists or institutions: according to one of Lang's theatre advisers, François Mitterrand thus insisted on one occasion that the *direction du Théâtre* provide funding for the mime artist Marcel Marceau, who had not previously received a subsidy from the Ministry.[9]

However, while ministerial and presidential interventions underlined the regal origins of public support for the French theatre, they did not modify the substance of the Ministry's funding policy. By contrast, as noted earlier, Lang did play a crucial role both in the near-doubling of his Ministry's budget in 1981, and in according a significant overall budgetary increase to the theatre sector, which made many of the individual decisions discussed here possible. In order to complete our discussion of the budgetary politics of the subsidized theatre, we now turn to the factors which influenced the setting of the overall theatre budget during Lang's tenure as Minister.

The Sectoral Budget: Ministerial and Professional Influences

The relative stability of the level of budgetary resources available to the *direction du Théâtre* after 1982 was no doubt explained in large part by Jack Lang's personal interest in, and attachment to, the theatre, and by his personal commitment to maintaining the sector's overall budget at its 1982 level. In the words of one of Lang's *cabinet* advisers: 'I saw Jack Lang call his counterpart at the Ministry of Finance several times to say, "Leave the theatre alone!"... Jack Lang's cultural policy may be debatable, but the theatre was a priority for him in the state apparatus.'[10] However, Lang's personal commitment to the sector coincided with another, less obvious source of pressure for stability in the theatre budget. Uniquely in its relations with different groups of performing arts professionals, the Ministry was faced, in the theatre sector, not only with influential individuals, but also with a highly disciplined and representative lobbying organization articulating the interests of the profession as a whole, the *Syndicat national des directeurs d'entreprises artistiques et culturelles* (SYNDEAC). The precise power resources of the SYNDEAC in the period studied are difficult to gauge, not least since there were no cases of open conflict between the Ministry and the theatre profession during Lang's tenure as Minister. Nonetheless, it is possible to make inferences about the SYNDEAC's impact on budgetary policy-making from the way in which individual officials evaluated the organization's influence. The assessment of Robert Abirached, *directeur du Théâtre* from 1981 to 1988, is worth quoting at length in this context:

> Lang loathed discussions in which he was disputed... . He would say, 'Take care of this for me, I don't want to see [the SYNDEAC]'. He didn't want any contact with them. So, as you will note – and there is no contradiction – for as long as Lang was there, he made sure that funding for the theatre did not decline too much. Because it's quite straightforward: if you look at the world of music, if you look at the world of dance, if you look at the world of museum curators, if you look at painters, the visual arts, there isn't even the beginning of anything resembling an organization. It's not the same thing. But the SYNDEAC only has to press a button, and within ten days, you have a petition. As they have money, because they bring together all the directors of the institutions, they're able to buy a whole page in a newspaper... And that's what a politician is most afraid of – it's the capacity to generate publicity, it's the organizational capacity, it's the capacity to mobilize.[11]

In addition to its confrontational capacities, which, as we noted, were never put to the test during Lang's tenure as Minister, the SYNDEAC played a more collaborative, but no less effective, role as the Ministry's privileged interlocutor in the routine administration of the sector. With regard to certain technical issues, such as the renegotiation, at the beginning of the second Lang Ministry, of the contracts specifying the artistic and financial obligations of the *centres dramatiques nationaux*, the SYNDEAC thus acted as the sole negotiating representative for all affected institutions. Moreover, the acceptance of the SYNDEAC as a legitimate and representative negotiating partner involved regular informal contacts in which the SYNDEAC was informed, not least, of the Ministry's financial plans for the sector at an early stage of the budgetary process. According to one senior SYNDEAC officer, the organization was thus able, if not to 'censure' the Ministry's projects, at the very least to impress its views about the Ministry's overall funding priorities on its official interlocutors:

> The relationship you have with the *direction du Théâtre* can influence the *direction du Théâtre* with regard to the allocation of the overall amounts of funding – 'Be careful with the CDN', 'Be careful with the independent companies', 'Be careful with the *théâtres nationaux*'. ... But as for more specific interventions – for example, saying to the *directeur du Théâtre*, 'Look here, this particular CDN is in difficulty' – it's something you can do, but it doesn't really have a direct influence, because the mobilization needed to ensure that one of us be treated justly, the collective mobilization doesn't exist. Not at all.[12]

Conversely, the Ministry was careful not to accord the SYNDEAC any substantive role in the allocation of subsidies, according to a senior official at the *direction du Théâtre*:

> There have always been discussions on this issue – that is, the SYNDEAC wanted to know more or less what the list of grants looked like – whom we'd cut, to whom we'd give more, etc. ... But to each his responsibility – and so there was no question of some sort of *cogestion* of the subsidies with the SYNDEAC. ... It's not the SYNDEAC's role to tell the state who needs to have their funding increased, who should be appointed, etc. Informally, of course, we would sound out the entire profession ...[13]

The SYNDEAC could thus be said to have exerted a strong influence with regard to the overall volume of funds allocated to the theatre, while its impact on the distribution of subsidies was much less significant. Again, this tends to underline the central point to emerge from our account of the SYNDEAC's role in theatre policy-making – that is, the crucial importance of the theatre profession's capacity for effective collective action, which was inevitably strongest in the area of overall budgetary allocations, and much more difficult to mobilize or sustain with regard to internal distributive issues. To further underline and elaborate on the importance, in the theatre sector, of this unusual propensity for collective action, we now move on to consider relations between the Ministry and the independent companies, an area in which the SYNDEAC, again, played an important role.

5.3 POLICY TOWARDS THE INDEPENDENT COMPANIES

As we noted in an earlier section, the dramatic expansion of funding for independent theatre companies was the most noticeable feature of the first theatre budget elaborated after Jack Lang's appointment as Minister of Culture, and the concomitant multiplication of the number of independent companies was perhaps the most important development in the sector as a whole in the 1980s. Thus, in the first year of Lang's tenure as Minister, the number of subsidized companies rose from 189 to 356; moreover, a substantial number of companies were promoted to permanent, or *hors commission*, funding status (see Figure 5.3).[14]

Particularly during the second Lang Ministry, however, the 'inflation' in the number of companies, which was not accompanied by a corresponding growth in audiences, became a point of criticism frequently levelled at the Ministry. Indeed, successive *directeurs du Théâtre* publicly suggested after Lang's re-appointment as Minister of Culture in 1988 that an outright reduction in the number of subsidized companies might be desirable. In 1989, the Minister himself conceded that 'more rigour, more coherence, more rationality' might be required in the Ministry's choices in this sub-sector.[15] However, the determined

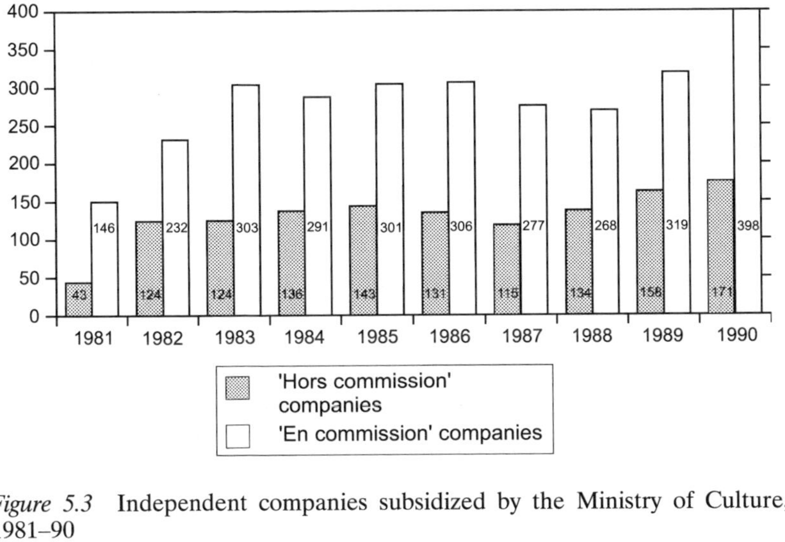

Figure 5.3 Independent companies subsidized by the Ministry of Culture, 1981–90

Source: *Ministère de la Culture*.

opposition of the professionals concerned – in the words of one of Jack Lang's *conseillers techniques*, the 'fierce *immobilisme*' of the profession[16] – would once again constrain the options open to policy-makers, and effectively render the expansion of the companies sub-sector irreversible. In 1988, Robert Abirached, when asked about the desirablity of reducing the number of companies supported by the Ministry, thus noted in a press interview:

> It's a legitimate question. But I can only observe that whenever I've wanted to attempt a reform to this effect, it provoked a general outcry of public opinion: the press, members of the profession, deputies, local departmental councillors. I've even received threats, anonymous phone calls at my home, at night.[17]

In order to explain both the initial decision to multiply funding to the independent companies, and the obstacles to the Ministry's projects for reform of the companies sub-sector, it is necessary to sketch briefly the historical evolution of the 'non-institutional' theatre in France. The emergence of the independent companies as an artistic and organizational phenomenon was closely linked to the 'events' of 1968, which had a strong 'politicizing' effect on French theatre professionals in general, and to the perceived inaccessibility, to a then emerging generation of directors and actors, of the 'institutional' theatre. Perhaps most significant, the *compagnies* conceived of themselves as belonging to a 'movement' that would articulate its demands collectively – unlike the institutional theatre, traditionally based on an exclusive and highly personalized relationship between the Ministry of Culture and a small number of privileged 'insiders' among the profession. This capacity for collective pressure politics would, throughout the Lang years, be the most important resource of the independent companies in their relationship with the Ministry of Culture.

The emergence of a collective lobbying strategy on the part of the independent companies predates the creation of the SYNDEAC: in 1971, a number of directors had established a representative organization, the *Action pour le Jeune Théâtre* (AJT), which articulated the companies' demand for a greater share of the theatre budget – with some success, as the budget of the commission responsible for attributing subsidies to the companies was more than quadrupled in the first few years of the AJT's existence. The AJT as such was dissolved in 1978, not least since many of its original members had moved into the 'institutional' sector, often as directors of CDN, and thus attained their principal material objective. However, the tendency for the independent companies to engage in an unusually vociferous brand of pressure politics continued in the 1980s. Increasingly, the SYNDEAC now became the official conduit for the companies' demands. In the 1970s, the chasm separating the companies and the 'established' theatre had been reflected in a strict division between the 'radical' AJT and the 'institutional' SYNDEAC, whose membership then consisted exclusively of the directors of large, permanently funded institutions. In the words of one of the AJT's founders,

> the SYNDEAC seemed, from the start, like a grouping of 'notables'. There were directors of *centres dramatiques*, who before 1975 were people who had already reached a certain age . . . , and so relations between the AJT and the SYNDEAC were necessarily a bit difficult.[18]

Although the two organizations occasionally co-ordinated their activities, both saw their interests *vis-à-vis* the Ministry as fundamentally diverging.

However, in the mid-1980s, the SYNDEAC's position towards the companies gradually evolved. The shift resulted partly from a generational change within the 'institutional' theatre, many of the directors appointed since the mid-1970s having themselves begun as directors of independent companies, and partly from a concerted effort by a number of companies to change the SYNDEAC's policy from within. Thus, a former president of the AJT, René Loyon, was elected to the SYNDEAC's National Council in 1985 and subsequently established a 'Commission Compagnies' within the SYNDEAC; this in turn encouraged an ever greater number of companies to join the organization. Indeed, by the end of the second Lang Ministry, the companies had become the numerically largest 'sub-sector' represented in the SYNDEAC. The 'entry' of a large number of companies into the SYNDEAC in the mid-1980s coincided with a significant change in the organization's leadership style: in 1986, the presidency of the SYNDEAC, then held by Jacques Blanc, who eschewed the politics of contestation advocated by many of the companies, was assumed by Denis Guénoun, himself formerly the director of an independent company. According to one director of an independent company and member of the SYNDEAC Council, this change reflected 'the fact that people were to a certain extent fed up with men like Blanc, with men who operated by backstage manoeuvring',[19] and entailed an important shift in the SYNDEAC's position, which contributed further to curtailing the options open to the *direction du Théâtre* in its policy towards the companies.

As we noted earlier, the Ministry shied away from directly confronting the issue of the companies during Jack Lang's tenure as Minister. The only instance in which an attempt was made overtly to reduce the number of subsidized companies occurred at the beginning of the first *cohabitation*, when Lang had already been replaced with François Léotard. Although we are concerned primarily with the policies pursued during the 1981–6 and 1988–93 legislatures, the attempted change of policy towards the companies under Léotard merits at least some attention, as it illustrates both the nature of the SYNDEAC's influence in the formation of policy and, more generally, the extreme susceptibility of policy-makers to open pressure from professionals in the theatre sector.

In the autumn of 1986, Robert Abirached, who had remained in his post as *directeur du Théâtre* after the change of government in March 1986, proposed a partial reform of the funding system, which accorded greater importance than previously to project-specific subsidies or *aides au projet*. As a consequence of Abirached's proposed reform, 28 *hors commission* companies which had previously been subsidized on a permanent basis would have been eligible only for *aides au projet*. As a member of the SYNDEAC Council recalls,

> not only the 28 companies concerned, but all of the companies reacted extremely badly to this – because we had the impression that the state all of a sudden was going to let go of its responsibilities … .We could sense a protest movement growing among the companies, but also, more generally, in the theatre world at large.[20]

The SYNDEAC organized a campaign of public protests against Abirached's proposed reform, which culminated in a demonstration in Paris in December 1986. As a result, the *direction du Théâtre* quickly retreated and withdrew its proposals. Concomitantly, the success of the protests persuaded a large number of companies to seek membership in the SYNDEAC, further reinforcing the influence of the companies sub-sector as a whole within the organization.

The threat of public protest by the SYNDEAC, the effectiveness of which was amply borne out by the protest campaign of 1986, was not the only external influence constraining the Ministry's policy *vis-à-vis* the companies. More discreet pressures were notably brought to bear on the Ministry by local politicians in support of particular companies. As one of Jack Lang's former advisers on theatre policy recalls,

> with regard to the list of companies [funded by the Ministry], every *directeur* [*du Théâtre*] arrives and says, well, it's necessary to make choices here. ... I confess that I would very much have liked to do some sorting in the list of companies. ... But on this list, there are companies... that one doesn't touch. There are political reasons for this.... And you don't touch this company or that because it's a company that's established in a Socialist town, and because the mayor will come to the defence to make sure you don't touch this particular company.[21]

Interventions by local politicians, in other words, further strengthened the obstacles to effective reform of the Ministry's funding policy towards the independent companies. By implication, this further contributed to the fundamental stability of budgetary distributions in the theatre sector after 1982, to which we alluded at the beginning of this chapter.

Before we conclude our account of policy-making in the theatre sector, one crucial facet of relations between the Ministry and its 'clients' remains to be considered. We noted earlier that the Minister of Culture had potentially far-reaching powers of appointment in the theatre sector: indirectly, these might be assumed to have given the Ministry substantial leverage over individual professionals and thus undermined the collective action capacity of the theatre profession as a whole. In order to test this proposition, and to reconcile the strong potential for a patronage-based system with the relative autonomy which characterized the actions of the profession in the cases discussed so far, we now turn to the factors which affected and constrained the actual use of the Minister's prerogatives.

5.4 APPOINTMENTS: THE *'FAIT DU PRINCE'* AND THE THEATRE

The extensive powers of appointment vested in the Minister of Culture and the President of the Republic reflect more strongly than any other facet of the formal theatre funding system the absolutist origins of public support for the theatre in France. In practice, however, in the period considered, a number of external influences and constraints worked to dilute the hierarchical nature of decision-making even with regard to this most 'monarchic' aspect of theatre policy-making.

Not surprisingly, a 'political' mode of filling senior posts in the subsidized theatre could be said to have applied most unambiguously at the top of the sector's institutional hierarchy – that is, in the case of the five *théâtres nationaux*, notably the *Comédie Française*. The *Administrateur*, or director, of the *Comédie Française* is appointed by prime ministerial decree in the Council of Ministers – a procedure that requires the co-signature of the President of the Republic. Allegiance to the governing or presidential party has consequently been a minimum prerequisite for appointment to this, the most prestigious post in the French theatre. In the period we are concerned with, the presidential prerogative was neglected somewhat as a result of François Mitterrand's personal lack of interest for the theatre as an artform, and appointments even at the level of the *Comédie Française* were largely delegated to the Minister of Culture. At the *Comédie Française*, the Giscardian Jacques Toja was thus replaced in 1983 by Jean-Pierre Vincent, previously director of the *Théâtre National de Strasbourg*, whose artistic credentials were undisputed within the profession, but who had also been a supporter of the Socialist Party since the late 1970s. The decision to appoint Vincent was taken by Lang, albeit on a recommendation from Robert Abirached, then *directeur du Théâtre*, rather than by the President. Similarly, in 1988, on the Socialists' return to power, Antoine Vitez, with whom Lang had collaborated at the *Théâtre National de Chaillot* in the early 1970s, was named *Administrateur*. In practice, then, the *Comédie Française* lay within a ministerial sphere of decision-making.

However, while Mitterrand showed little interest in the substance of theatre policy, he did reserve the right, as in other policy areas, to impose presidential nominees for particular posts regardless of the preferences expressed by the Minister of Culture, or of the reactions such interventions might provoke among the theatre profession. Most controversially, Mitterrand decided in 1988 to impose the director Jérôme Savary, who had in the past assisted the President with his election campaigns, as head of the *Théâtre National de Chaillot*.[22] Savary's appointment provoked fierce criticism within the profession; perhaps more important, it explicitly bypassed the *directeur du Théâtre*, Robert Abirached, who was informed of the decision only retrospectively, and who resigned soon afterwards. However, the circumstances of Savary's appointment were exceptional: while Mitterrand's intervention served to underline the extent of presidential power, the hostile reactions provoked by the appointment both within the administration and among the profession reflected the strength of a more consensual tradition of decision-making.

At the level of the *centres dramatiques nationaux* and the *maisons de la culture*, appointments were formally subject to ministerial decision-making. In practice, Jack Lang generally devolved responsibility to his *directeur du Théâtre*, while reserving the right to impose a more personal choice in particularly sensitive cases. Perhaps the most important example of ministerial intervention in this area was the transformation, in 1981, of the financially troubled *maison de la culture* at Nanterre into a *centre dramatique national*. The institutional change was closely bound up with the appointment of a suitable director for the new institution. While Patrice Chéreau, the new director, had already been mooted as a candidate for Nanterre before 1981, Lang's active support for Chéreau's candidature was decisive in overcoming the resistance mounted by the personnel of the old *maison de la culture* and by the Communist municipality of Nanterre. According to one of the administrators of the *Théâtre des Amandiers* created by Chéreau in the buildings of the old *maison de la culture*,

> one has to recognize that the combined determination of the Minister and of the then *directeur du Théâtre* really was decisive. Because even with someone like Patrice Chéreau involved, it was very difficult, in 1981–82, to get rid of a *maison de la culture* . . . , it was far from obvious that they would succeed.[23]

In Lang's own words, 'it was such a disaster that I took charge of the case myself, and I said, "Stop it, we're winding up, we're abolishing [the *maison de la culture*], and we're appointing Chéreau." '[24]

However, more routine appointments at the level of the *centres dramatiques nationaux* were generally considered the preserve of the *directeur du Théâtre*. According to Robert Abirached, Lang intervened only exceptionally:

> There were two or three cases. For example, he wanted to change the directors of two *centres dramatiques nationaux*, I don't know why, he almost ordered me to do it, and I did it. I didn't understand the reasons, but afterwards I had to deal with the problems. Because I had told him that there would be a lot of problems, which he got – that is, even interventions coming from the Presidency of the Republic, because they were well connected. So you had this, you had one or two cases of favours being asked for particular individuals, but this is something you're obliged to do, it's a tiny fragment – it's

nothing when it represents only 0.5 per cent of your decisions, you even congratulate yourself when that's all it is.[25]

Regardless of the precise distribution of competences between the Minister and his *directeur du Théâtre*, however, another factor, external to the Ministry, increasingly came to bear on the the appointment process. By the end of the 1980s, sub-national government, notably the communes and regions, provided around one third of the total subsidy for the CDN and an even more substantial proportion of the income of the *maisons de la culture* and *centres d'action culturelle*. Consequently, according to a member of Lang's *cabinet*, 'at the level of the *collectivités locales*, it's important nowadays – a bit too important, some would say – that if you're appointing someone in a region, you'd best avoid a wrestling match with the municipality or with the region.'[26] Thus, while the Minister of Culture and his *directeur du Théâtre* retained a pre-eminent position in the appointment process, local policy-makers were increasingly able to exercise, at the very least, a right of veto over appointments to institutions funded jointly by the Ministry and the *collectivités locales*. In some instances this necessitated protracted negotiations before a candidate could be agreed upon. As in the case of the independent companies, the prerogatives of national policy-makers were thus increasingly constrained by the need to accommodate sub-national influences.

To summarize the above, appointments in the subsidized theatre were subject to a much wider range of influences than we might suspect simply by considering the formal structure of the funding system. Both the Minister and the President of the Republic occasionally made use of their powers to impose their preferred candidates for particular institutions. However, in most cases, appointments were less politicized and reflected an underlying consensus between the Minister, the administration and the profession about the desirability of preventing an explicit politicization of the appointment process. This, in turn, diminished the Ministry's ability to use appointments systematically as a source of political patronage and as a mechanism for exercising control over the profession, a fact that further contributes to reconciling the apparent domination of the sector by the state with the *de facto* influence of theatre professionals.

CONCLUSION

Throughout this chapter, our discussion has revolved around a paradox. Of all the performing arts sectors considered in this book, the subsidized theatre conforms most closely, in its institutional structure, to a statist cultural policy paradigm. Both the rigid stratification of the sector and the Minister's far-reaching powers of appointment with regard to the most important theatrical institutions in France would suggest a hierarchical, or clientelist, relationship between the Ministry and the professionals subject to its control. In practice, however, we found that hierarchical policy-making, in a ministerial, presidential or bureaucratic guise, was effectively confined to the margins of theatre policy. By contrast, the impact on policy-making – or, conversely, the ability to block structural changes – of the Ministry's professional 'clients' was exceptionally pronounced. In order to account for this gap between institutional appearances and actual policy-making patterns, we focused on the resources on which the theatre profession was able to draw in its relations with the Minister and civil servants. Throughout our analysis, the exceptional organizational capacity of the theatre profession recurred as the single most powerful factor constraining and shaping the policy-making process, and overriding the statist form of the funding system. The ability of very disparate sections of the profession to cohere around certain central issues and pursue an effective collective action strategy *via* the theatre directors' union, the SYNDEAC, had very tangible implications for the choices made by the Minister and successive *directeurs du Théâtre*, notably in the area of budgetary policy. In the case of the independent companies, similarly, we found that the Ministry's policies were conditioned by the pressure exerted by the profession, notably *via* the SYNDEAC; in 1986, the attempt by the incoming conservative Minister to revise the status of a large number of companies led, exceptionally, to open conflict – the outcome of which confirmed the efficacy of the collective action strategy adopted by the profession. In a comparative perspective, the exceptional internal cohesion of the theatre profession, reflected in the unusually representative character of the SYNDEAC, is perhaps best explained by the strength of a certain ideological tradition, moulded by the 'events' of 1968, among the generation of directors which occupied the central institutional positions in the period we studied – and which was unique, among the sectors considered in this book, to the theatre.

In short, then, we found that the resources and strategies of the Ministry's major professional 'clients' imposed constraints on policy-makers which effectively negated the formal institutional capacities for statist policy-making inherent in the structure of the funding system. However, while a focus on the resources of the profession provides the core of our account of policy-making in the theatre sector, the policy network approach, introduced in an earlier chapter as the centrepiece of our analytical framework, should by no means be seen as redundant. Throughout the period studied, relations between the Ministry's *direction du Théâtre* and a core 'clientèle' of established professionals were marked not by overt or latent confrontation, but by 'complicity'[27] and a strong consensus about the goals and priorities of theatre policy, corresponding to the 'policy community' ideal type defined by Rhodes and Marsh. It should be stressed that policy network analysis on its own does not provide a satisfactory characterization of the policy-making process, notably with regard to policy towards the more marginal elements of the sector, such as the independent companies. That is, while relations between the Ministry and its 'clients' in the companies sub-sector conformed closely to the issue network paradigm, policy outcomes diverged significantly from the patterns associated with issue networks in the Rhodes and Marsh framework. The reasons for this hiatus in a network-based account of the relationship between patterns of policy-making and policy outputs were of course related to the exceptional strength of the collective action strategies deployed by theatre professionals. To a much greater extent than in other sectors, the policy network approach should thus be seen as complementary to, rather than as a substitute for, explanations centred on the power resources of individual actors. The issue of establishing a balance between network analysis and agency-centred explanations should not, however, obscure the central theme of this chapter – that is, the limitations and constraints placed on hierarchical decision-making even in a sector which formally epitomizes the French tradition of active state intervention in the performing arts.

6 Film Policy

The sectors considered in previous chapters are all characterized by a similar underlying economic structure: the Ministry's 'clients' in the performing arts are without exception non-profit making and dependent for their survival on extensive revenue-funding by government. The cinema presents rather different economic characteristics. The production, distribution and exhibition of films are, in France as in other countries, commercial activities; indeed, a number of the most important producers in the sector are linked to some of France's largest industrial and financial concerns. Moreover, the sector is exposed to severe competition both from foreign products and from alternative media – the issue dominating relations between government and the industry is thus not simply one of financing deficits, as in other sectors, but one of protection against external threats.

These peculiarities might suggest that any comparison with such sectors as the theatre or contemporary music will be a precarious exercise. In fact, as we will attempt to show in this chapter, focusing on three specific policy initiatives aimed at film production, there are strong parallels, where policy formulation is concerned, with the other sectors considered. The central issue, against which the narrative in each case study has to be read, will, as in previous chapters, be the relationship between the core executive, or its individual components, and its private-sector 'clients'. While there are significant differences between policy-making styles in the respective case studies, we will argue that a clear pattern can be detected, which contributes to the emerging picture of a generic policy style applicable to cultural policy as a whole.

6.1 FILM POLICY, 1981–93: CONTINUITY AND CHANGE

The focus of this chapter is, of course, on policy-making in the film production sector during the 'Lang years'. However, in order to understand both the policies initiated under Jack Lang and the policy-making process, it will be necessary to outline some of the structural constraints and influences that conditioned the choices of policy-makers in the 1980s and 1990s. The following section will thus, first, provide a general overview of the evolution of public support for the cinema; second, summarize – though by no means exhaustively – the policy of the Ministry towards film production during Lang's tenure as Minister; and, finally, describe the organizational structure of the cinema 'lobby' as it evolved in the 1980s. We will then be able to consider the policy process with regard to three specific measures introduced under Jack Lang.

The French Film Industry: Surviving against the Odds

In France, as elsewhere, the audience for cinema has diminished dramatically as a direct consequence of the progress of television. From 343 million seats sold in 1965, attendance fell to just 130 million in 1987. The total capacity of French cinemas, too, has been in decline since the 1960s, and continued to fall throughout the 1980s, from over 1.3 million seats in 1982 to less than one million in 1992. Perhaps most important, the market share of French films has been continually eroded on their home ground in the face of American competition: in 1994, the share of French films fell below 30 per cent of seats sold for the first time, from around 50 per cent in 1981.[1] However, the French film industry has been more resilient than its counterparts in neighbouring countries – since the early 1960s, cinema attendances in the United Kingdom, Italy and Germany have declined from levels substantially higher than those in France to levels far below those in France. Moreover, France stands out from its neighbours in that French films retained, throughout the 1980s, a falling but nonetheless significant share of their home market. Concomitantly, the French film industry has maintained the infrastructure necessary for large-scale production; in 1993, investment in domestic film production totalled around 2.3 billion francs. The structural core of the production sector has remained intact, centred on a small number of vertically integrated *groupes* – Gaumont, UGC and (since its acquisition by the Chargeurs conglomerate) Pathé – which dominate cinema ownership and play a preponderant role in production. At the same time, a large number of independent producers have survived. Within the industry, it is possible to discern a number of important cleavages – between the *groupes* and the independents, between production and exhibition interests – which have had important implications for the structure of the public funding system, and we will return to these later.

Sources of 'Public' Finance for the Cinema: from Vichy to Lang

Since 1946, public policy towards the cinema has been administered by the *Centre national de la cinématographie* (CNC). Though formally an administrative *établissement public* attached to the Ministry of Culture, the CNC effectively operates as a directorate of the Ministry, similar to the *direction de la Musique et de la Danse* or the *direction du Théâtre et des Spectacles*, its Director-General being appointed by, and answerable to, the Minister of Culture like any other sectoral *directeur*. The CNC – the creation of which pre-dates that of the Ministry itself – is also marked by the legacy of the *Comité d'organisation de l'industrie cinématographique* (COIC), a body established by the Vichy régime to oversee the industry, and from which the CNC inherited many of its far-reaching regulatory powers. At the same time, the CNC assumed the 'corporatist' aspect of the COIC – more than any of the traditional ministerial directorates, which cling, at least formally, to the discourse of 'impartiality' that has traditionally characterized the French administration, the CNC has close institutional ties with its 'clients'.

The closeness between the CNC and the industry is explained, in large part, by the peculiar system by which public subsidy for the cinema is raised and allocated. Given the technical complexity of the system that has evolved since the inception of the CNC, we will consider only the most important elements of the system, and focus on support for production (rather than distribution or exhibition, which are also subsidized by the CNC). Since 1959, the various support mechanisms administered by the CNC have been funded almost exclusively by a levy imposed on the sale of cinema tickets, the *Taxe Spéciale Additionnelle* or TSA, amounting to around 11 per cent of box office revenue. The TSA is paid into a special fund, the *compte de soutien*, administered by the CNC,[2] which allocates funding between different support mechanisms. By contrast, the volume of budgetary subsidies as such has remained comparatively low – Farchy (1992, p. 213) speaks of 'modest budgetary contributions which have been obscured by the existence of the *compte de soutien*'. There have, however, been important variations in the level of direct subsidy, particularly when François Léotard, during the first *cohabitation*, sought to progressively phase out the direct, budgetary component in cinema funding. Léotard's policy was reversed after the return of Jack Lang in 1988, when direct subsidies were increased to more than their pre-1986 level (see Figure 6.1).

Since 1986, the *compte de soutien* has derived a large proportion of its revenue from a special tax on television companies – by 1991, the networks' contribution to the *compte de soutien* amounted to 985 million francs, more than twice the amount raised by the TSA. The imposition of a tax on the TV companies coincided with the creation of

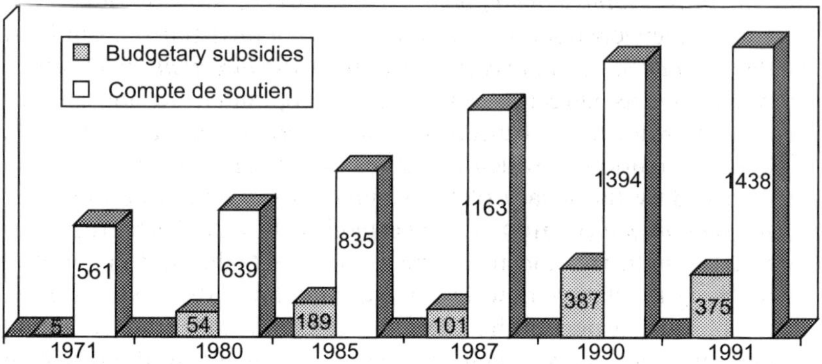

Figure 6.1 Public support for cinema and audiovisual production: budgetary subsidies and *compte de soutien*, 1971–91 (*dépenses ordinaires* and *autorisations de programme*, millions of francs indexed to 1991)

Source: *Ministère de la Culture*.

an 'audiovisual' section in the *compte de soutien*, such that the audiovisual sector partly recovered the tax in the form of production subsidies. However, around 40 per cent of the income generated by the tax was allocated to the 'cinema' section of the *compte*, making the system strongly redistributive. We will examine the tension inherent in this arrangement, and more specifically, the decision to impose a special tax on the networks, in greater detail in a subsequent section of this chapter.

The funding mechanisms supported by the *compte de soutien* fall into two categories – 'automatic' and 'selective'. Of the former, the *soutien automatique* for the production of feature-length films (*longs métrages*) is by far the most important in financial terms (see Figure 6.2). The *soutien automatique* for a particular film is based on the reimbursement to producers of a fixed proportion of the TSA collected at the film's showing in cinemas. Until 1989, the proportion reimbursed varied according to the total revenue a film generated; since 1989, a single rate of 120 per cent of the TSA collected has been applied. Only French films are eligible for the *soutien automatique* – as all cinema tickets are subject to the TSA, the system is thus heavily subsidized by the box office receipts of American films. In addition to the *soutien automatique*, producers can, under the selective *avance sur recettes* scheme, apply for interest-free loans which are repayable only if a film generates sufficient box office income after release. The

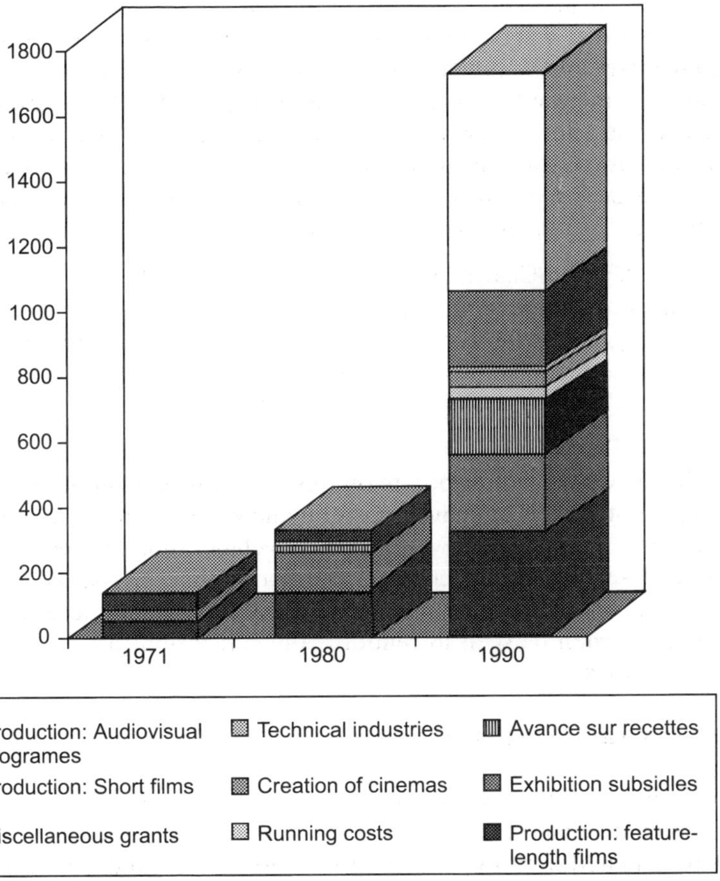

Figure 6.2 Breakdown of public funding for cinema and audiovisual production by beneficiary: direct subsidy and *compte de soutien*, 1971–90 (millions of francs)

Source: *Ministère de la Culture.*

avance sur recettes is attributed by a commission, appointed by the Minister, but on which the main industry interests are well represented. Similar selective support mechanisms, based on panels of professionals, exist for virtually all other branches of the film sector, underlining the extent to which the industry is formally implicated in funding decisions.

The compte de soutien: public money or private money?

The fact that direct budgetary subsidies account for only a very small percentage of the funds allocated by the CNC has important implications for the way in which state support is perceived by the profession. Thus, the film industry has traditionally considered the *compte de soutien* as constituting not so much a form of public subsidy, but rather a fund sustained by the industry itself and thus at the industry's disposal. As a member of Lang's *cabinet* recalls,

> one day, the Chairman of one of the big French corporations, who happened to be in my office, told me, ... 'The *compte de soutien* is money that belongs to the profession', and I told him, more or less to provoke him, 'No, it's money that belongs to the state.' And he said: 'If that's how it is, I'm leaving your office. I'm walking out.' ... So it was that kind of attitude.[3]

While we will not elaborate on this revealing ambiguity surrounding control of what is effectively a tax on audiences, it should nevertheless be borne in mind when we consider the elaboration of specific policies in later sections of this chapter. The notion that the *compte de soutien* 'belongs' to the industry, a view strongly held among leading producers, will then be seen to reinforce the conclusions arising out of the case studies.

The Cinema Policy of Jack Lang

The core of the system which had evolved since 1946 not only remained intact throughout Lang's tenure as Minister, but was strengthened both by an overall increase in the funds at its disposal, and by a number of new initiatives. The most significant changes implemented during Lang's tenure inevitably concerned the relationship between cinema and television – as already mentioned, Lang instituted a tax on the resources of television companies which revolutionized the financing of the *compte de soutien*. Moreover, the first Lang Ministry saw the creation of the pay-TV station Canal Plus, requiring severe regulatory arrangements to attenuate the threat the new station posed to the film industry.

With regard to the structure of the film industry as such, the initial impulse following Lang's arrival at the Ministry was to pursue a 'radical' policy in keeping both with the cultural agenda developed by the Socialist Party in the 1970s and with Lang's own public pronouncements on the desirability of curtailing the power of cultural 'monopolies'. Thus, one of the Minister's first decisions in the film sector, following on from a report commissioned by Lang from Jean-Denis Bredin, a prominent lawyer and friend of Lang's, was the dissolution of the Gaumont-Pathé distribution cartel. The move was welcomed by independent exhibitors, who had been pressing Lang to act against what they saw as a monopoly threatening their very survival, and it was vigorously opposed by Nicolas Seydoux, owner of Gaumont.[4] The difficulties of imposing the decision against Gaumont-Pathé are underlined by Lang himself:

> It was very difficult. And in order that the Gaumont-Pathé group did not see it as a kind of brutal, unbearable violation, I met them repeatedly, to tell them that if I did not decide a minimum amount of measures to limit concentration, then there was a risk that the then parliamentary majority, the very purist Socialist majority, might pass much more severe measures in the National Assembly.[5]

However, the breaking-up of Gaumont-Pathé did not lead to a more systematic anti-trust policy. While the Ministry did increase funding for independents and regional cinema, the 1980s were also marked by increasing concentration within the industry, and by closer ties between the film sector and large industrial and financial concerns. UGC, one of France's two established integrated groups, alongside Gaumont, had long been controlled by a group of large institutional shareholders, led by the Compagnie Générale des Eaux. The Pathé chain of cinemas was taken over in 1989 by the Chargeurs corporation – controlled by Jérôme Seydoux, brother of the Gaumont chairman – which subsequently diversified into film production. More significantly, a *rapprochement* took place, in 1992, between Pathé and Gaumont, with the two groups effectively establishing a geographically co-ordinated exhibition cartel.[6] By now, Lang no longer endeavoured to control the large groups – in any case, in the words of one of France's most influential producers, 'given the weight that the Seydoux family has in the French cinema, it's very difficult to impose on them a reform that they don't want'.[7] Moreover, many of the Lang Ministry's most decisive measures aimed at stimulating production, while not targeted explicitly at the *groupes*, benefited the large concerns disproportionately and reinforced the presence of the financial sector in the film industry. Among these, the creation of institutions intended to encourage private investment in film production – notably the *Sociétés pour le Financement de l'Industrie Cinématographique et Audiovisuelle* (SOFICA) and the *Institut pour le Financement du Cinéma et des*

Industries Culturelles (IFCIC) – had the greatest impact. The Lang Ministry – specifically, in the guise of the 'plan Lang' of 1989 – also introduced a number of changes in the financial rules governing the attribution of the *soutien automatique*, such as to favour big-budget productions. Again, the *groupes* and the larger independents were the main beneficiaries.

Organized Interest Groups in the Film Sector: 'Le Lobby'

Before concluding this section, it will be useful briefly to adumbrate the structure of what is commonly referred to by policy-makers in the sector as 'le lobby', as this will play an important role in our discussion of specific measures introduced during the 1981–93 period. The professional organizations or *syndicats* that represent each of the various branches of the industry (production, distribution and exhibition) constitute the oldest example of organized private sector lobbying in the cultural policy arena, and virtually all important companies and individuals in the sector belong to one of these *syndicats* – although some of the most influential members of the *syndicats* are also, in the words of a senior Ministry official, 'individuals who are by themselves lobbies, powerful people'.[8] In this chapter, we will focus only on the producers' organizations – not least since our case studies deal exclusively with measures aimed at production, but also because, throughout the 1980s, the producers' organizations progressively increased their influence, becoming the politically most active and effective of the cinema *syndicats*.

At the beginning of the 1980s, there were two production *syndicats*, each with a clearly defined clientele. The *Chambre Syndicale des Producteurs et Exportateurs de Films Français* was seen to articulate the interests of the *groupes* – Gaumont foremost among them – and the most financially successful of the independent producers. The *Association Française des Producteurs de Films* (AFPF), on the other hand, which had been created by 'dissident' members of the *Chambre Syndicale*, represented smaller independent producers. As a function of their respective constituents' interests, the *Chambre Syndicale* has generally adopted a relatively market-oriented stance (without, of course, questioning the basic structure of the funding system) and thus advocated as extensive a use of *soutien automatique* as possible; the AFPF, conversely, has defended selective subsidies such as the *avance sur recettes* and argued, unsuccessfully, for anti-trust measures to deconcentrate ownership in the sector. This simple dichotomy between the 'big' producers and the 'small' producers was troubled somewhat when, in 1987, a group of producers split off from the *Chambre Syndicale* to form a new *syndicat*, the *Union des Producteurs de Films* (UPF). Ostensibly, the split was caused by personal rivalries between leading members of the *Chambre Syndicale*, and the two organizations had no significant policy differences. However, while the UPF's first president, Alain Terzian, was an independent, the UGC group played an instrumental role in the split. Moreover, the underlying reasons for the secession were indirectly related to the battle for the privatization of the TF1 television station, in which the *Chambre Syndicale* implicitly sided with the Bouygues group against another candidate, Hachette, which was supported by UGC. The *Chambre Syndicale* retained, as its most influential members, Nicolas Seydoux's Gaumont group and the producer-director Claude Berri (who would later establish a close partnership with the cinema subsidiary of Jérôme Seydoux's Chargeurs conglomerate). The UPF included among its members UGC and a number of influential independents like Marin Karmitz and Daniel Toscan du Plantier. In 1989, yet another *syndicat* was created – the *Société des Auteurs, Réalisateurs, Producteurs* (ARP). Ostensibly, the new organization was to defend the specific interests of the growing number of directors who produced their own films, and it attracted a large number of France's best-known directors. However, the ARP's actual role and *raison d'être* could also be seen as a function of its ties with the *Chambre Syndicale* – its founder, Claude Berri, remained an influential member of the *Chambre Syndicale*, and the Secretary-General of the *Chambre Syndicale*, Pascal Rogard, was also appointed Administrator of the ARP, thus becoming the most influential professional lobbyist in the industry. The rationale behind this move is summed up by a senior Ministry official: 'By creating the ARP, they acquired an instrument ... for saying, "But we're not simply people who defend Gaumont and the big producers, we've also got serious film directors." ... Those are tactical manoeuvres.'[9]

We will discuss the actual influence of the groups that make up the 'lobby' in greater detail with regard to the case studies considered below. However, we have dwelt on their role because the existence of highly organized and professional interest groups, and the stability of the positions they articulate, is a phenomenon which sets the French film industry apart from the other sectors considered in this book. It has to be seen, in other words, as an integral part of the structural context in which film policy was formulated during the 'Lang years' –

and one which would both condition and constrain that policy in important respects.

In the remainder of this chapter, we will consider the policy process with regard to three of the measures referred to above: the creation of the SOFICA; the regulation of television, to the extent that it concerned the film sector; and the support for big budget films initiated by the 'plan Lang' of 1989. In a concluding section, we will then consider what we can infer, from these case studies, about the relationship, and distribution of roles, between public and private actors in the film production sector.

6.2 THE CREATION OF THE SOFICA

In 1985, the Ministry of Culture created a new mechanism for enticing private capital into film production – the *Sociétés pour le Financement de l'Industrie Cinématographique et Audiovisuelle* (SOFICA). A law voted by parliament, and promulgated on 11 July 1985, established the framework for a new form of 'tax shelter' applicable exclusively to investment in film production. Thus, the sums invested in a SOFICA (a fund which could theoretically be set up by any individual or institution) would be tax-deductible for the investor.[10] In return, the SOFICA was obliged to invest the whole of its capital in the production of French film or audiovisual works; the composition of the SOFICA, and its investments, were to be subject to rigorous public supervision. The impact of the SOFICA on the sector has, in quantitative terms, been considerable. In 1991, 18 SOFICA were in operation. Between 1985 and 1993, their contribution to the total financial investment in French film production is estimated to have amounted to around 5 per cent; their share in the production budgets of the films they co-financed amounted to around 10 per cent on average.[11] However, while the SOFICA have been successful in widening the capital base for film production, some observers have argued that the mechanism has evolved in a way at odds with its original purpose of mobilizing venture capital from outside the industry. Thus, while the SOFICA were initially intended, at least in part, to encourage independent production, the major integrated groups, together with large financial institutions, gradually came to dominate the system, both because they had easier access to capital and because they were able to offer their investors better reimbursement conditions. Farchy (1992, p. 205) suggests that the SOFICA 'can seem like a hidden way of financing these companies' films at the taxpayer's expense, because the state bears most of the cost *via* the tax exemption mechanism'; at the same time, according to Farchy, they have essentially become

> a means – largely controlled by the banks that offer and manage them – of generating future audiovisual rights without any risk. Many of them have invested in films co-produced, or bought in advance, by the television stations, in order to secure the broadcasting revenues for themselves. This has been to the detriment of other films in which the TV stations are not involved.

Similarly, Racine (1994, p. 19) notes that 'a number of film production groups have understood that it is advantageous for them to create a SOFICA with the help of their usual bank or banks ...: it is a way of financing films at zero rate.' By the beginning of the 1990s, the SOFICA scheme had thus progressively moved further away from its stated aim, and from the intentions of its creators.

The Creation of the SOFICA: Public Sector Voluntarism Meets Private Sector Activism

The process that led to the establishment of the SOFICA system has to be seen, first of all, in the context of the profusion of new initiatives that characterized the first years of Lang's tenure at the Ministry of Culture. In an 'expansionary' phase that lasted until 1985, Lang, his *cabinet* and senior officials in the Ministry were actively casting around for innovative ways of stimulating cultural output in all domains that fell into the Ministry's remit. The cinema was undoubtedly one of the sectors – along with the new sectors of 'popular' culture like rock and *variétés*, which we will consider in a later chapter – to which Lang was personally most committed in this period. There was, in the words of one CNC official, 'a very strong ministerial impetus – Jack Lang really knew how to take on board the problems and struggles of the cinema professionals and make them his own. And in that sense he was a Minister who was close to them, and who really fought to get a number of things done.'[12] The unprecedented activism of the Minister and his administration coincided, in the early 1980s, with a growing awareness on the part of both officials and the industry of the financial difficulties facing film production – the rise in new media, coupled with a crisis in the traditional financing system based on distributors' advances, had increasingly led producers to lobby the Ministry for new sources of finance. As often in the sector, the concerns of the industry were quickly adopted by the Ministry and the CNC: in the words of one of Lang's advisers, 'the analysis was shared by everyone ..., but ... the members of the profession played an important role in drawing attention to certain issues, in explaining, in alerting the state.'[13]

From 1983 onwards, various solutions to the crisis in film financing were mooted in informal meetings between leading professionals and the small group of officials that could be said to have formed the 'core' of the decision-making nexus in this sector: the Director-General and Deputy Director-General of the CNC; the *conseiller technique*, or sectoral adviser, responsible for cinema in the Minister's *cabinet*; and, not least, the Minister himself, who had frequent personal contact with representatives of the industry. A large number of possible remedies to the dearth of capital in film production were suggested by all sides; these focused mainly on the role of the financial sector, and on ways of facilitating the access of producers to commercial credit. However, the idea that quickly gained most favour envisaged using some form of fiscal incentive to entice private investors into film production. 'Tax shelters' for film production had been experimented with in a number of countries, such as Canada and Australia; in France itself, the financing of film production by means of public subscription had already been tried in the 1930s, though never on a large, institutional scale. Moreover, the notion of using tax shelters to further specific public investment goals had been gaining ground in other sectors of the French administration. However, it was also recognized that many of the systems tried in other countries had been subject to abuse in the absence of adequate controls. The option of allowing individuals to invest directly in a particular film was thus rejected in favour of a system by which an intermediary, the SOFICA, would channel the funds invested in it by investors, subject to a licence accorded by the state.

It is difficult to pinpoint the precise origins, or reconstitute the 'directionality', of the formula finally adopted, and which in early 1985 was ready for submission to parliament as a government bill. Those involved agree that the SOFICA was the result of 'a kind of collective creativity', where 'everything was analysed jointly by those concerned and everything was decided jointly by those concerned'.[14] However, it is possible to differentiate between different groups of actors at different stages of the project's gestation. The initial impulse came from an independent producer, Jacques Perrin, who had experienced difficulties in raising finance for a project and thus, in meetings with CNC and Ministry officials at various levels, suggested setting up a scheme that would, with the help of fiscal incentives, channel private investment into film production. In the words of one CNC official, '[Perrin] started thinking about the issues, he even, in part, got the debate going and followed it at first, and then, afterwards, the system moved beyond his reach – the procedure, the mechanism, the administrative networks.'[15] Indeed, independent producers like Perrin looked at the time to be the main beneficiaries of a scheme that would greatly diversify the sources of investment capital. The system also promised to benefit the large corporations, however: in the words of a senior CNC official,

> it was also pushed by the *groupes* – Gaumont, UGC, Pathé and the others – because they saw in it a way of creating structures associated with them, in which they would involve the financial institutions, the banks, their own financing mechanisms .... So everyone was ultimately looking for a solution like this.[16]

In short, there was unanimous support among the profession for a 'tax shelter' system of the type sketched – crucially, the system would in no way affect the balance between the different constituent elements of the industry. It is worth noting, moreover, that the professional *syndicats* played only a subordinate role in the formulation of the SOFICA system: the scheme was elaborated by the *cabinet* and CNC officials in direct collaboration with selected professionals.

The project was strongly supported, then, by the industry, by the CNC administration, and by the Minister and his *cabinet*. The most difficult stage of the project's preparatory, pre-parliamentary phase – and one that was handled almost exclusively by Lang's *cabinet*, and notably the adviser responsible for cinema, Frédérique Bredin[17] – concerned its acceptance by crucial elements of the core executive outside the Ministry of Culture. Unlike policies 'internal' to the sector, such as measures affecting the distribution of the *compte de soutien*, the SOFICA had important fiscal repercussions. The measure thus impinged on the prerogatives of the Ministry of Finance, and particularly on the jurisdiction of the tax directorate, the *direction générale des Impôts*. Senior officials at the Ministry of Finance were ferociously opposed to the SOFICA when the system was first presented in meetings between representatives of the respective *cabinets* of Lang and the Finance Minister, Pierre Bérégovoy. Bérégovoy's *cabinet* relayed the 'functional' opposition of the financial administration – which was both concerned about a substantial loss in revenues and anxious not to set a precedent which might open the floodgates to similar schemes from other ministries. Thus, in the words of a senior member of Lang's *cabinet*, foremost among those resisting the measure from within the Ministry of Finance

> there were the tax directorate and the budget directorate, essentially, who were the ones with the most clout. Then there was also the department of fiscal legislation, which was our interlocutor at the time, and there was the Minister's *cabinet* – which is always a product of the Ministry's directorates, and which was of course extremely hostile to the system in question.[18]

The implacable opposition of the Ministry of Finance called for a political *arbitrage* at a higher level. Lang thus appealed to the Prime Minister, Laurent Fabius. Simultaneously, Lang's *cabinet* liaised with the Matignon staff. According to a senior CNC official, the fact that Frédérique Bredin's future husband, Jean-Pascal Beaufret, was a *conseiller technique* in Fabius' *cabinet* was an important factor in gaining the Prime Minister's support:

> There was a conjunction of individuals – which may be anecdotal, but it played its role. And there was a political conjunction: Lang and Fabius definitely got on rather well together. Then there was a second political conjunction, which was that the President of the Republic supported both Lang and Fabius. So the general circumstances were doubtless favourable. When the circumstances are favourable, the representatives of the cinema are very skilful at exploiting them to the utmost.[19]

Once Fabius had ruled in Lang's favour, against the Ministry of Finance, the technical details of the system could be elaborated. The relevant departments of the Ministry of Finance, notably the department of fiscal legislation, the *service de la législation fiscale*, became closely involved in drafting the specifications of the future bill, despite their prior opposition. Again, a personal factor played a crucial role in moving the project closer to implementation: Frédérique Bredin, who had become the project's most determined advocate inside the Ministry of Culture, belonged to the *Inspection des Finances* and was thus able to negotiate with officials at the Ministry of Finance on an equal footing. The CNC, by contrast, was not, at this stage, in a position to hold its own against the Ministry of Finance where the fiscal aspects of the project were concerned, as one CNC official involved in the discussions recalls: 'From the point of view of our fiscal competence, we weren't up to it. ... There wasn't a single person at the CNC who was on top of all that.' Moreover, the CNC could scarce presume to be taken as an equal by the Ministry of Finance in the informal status hierarchy within the administration: 'The Ministry of Finance is the most powerful bureaucracy in France – we at the CNC, we don't even exist, we're at best a bunch of Polynesians in the eyes of the Ministry of Finance.'[20]

After Fabius' ruling, the one important obstacle facing the project had been cleared; and Matignon having, in the words of a CNC official, 'put its foot on the accelerator',[21] the parliamentary phase of the process could be accomplished in a very short period of time. The SOFICA bill was presented to parliament in the spring of 1985; it became law on 11 July 1985 and was followed, on 17 September 1985, by two decrees establishing the technical framework for the creation of the SOFICA.

To summarize the above: the process of elaboration that ultimately led to the creation of the SOFICA law was initiated by individual producers concerned about the lack of private investment in film production, and who were seeking ways of attracting individual investors into the industry. Initially, these demands were expressed at the level of, and relayed by, the CNC, acting in its traditional role of 'interface' between the Ministry and the profession. However, the *cabinet* of the Minister of Culture, which was also in direct contact with representatives of the profession, played an increasingly important role as the project took shape. The role of the *cabinet* – and more specifically, that of Lang's adviser on film policy, Frédérique Bredin – became central once the process had become inter-ministerial, and once it had become clear that the determined opposition of the Ministry of Finance would require prime ministerial arbitration. At this stage, governmental – that is, prime ministerial – support became necessary to surmount the opposition to the project. It was at this stage, also, that ministerial intervention – Lang's appeal to Fabius – proved decisive. The project's elaboration could be attributed, then, to a genuinely co-operative effort involving members of a film policy network composed of selected professionals and administrators. However, the formal adoption of the project was possible only following an explicit ministerial decision and after the Minister had personally defended the project at the level of the government. Moreover, the system's main beneficiaries – the *groupes* – were far from playing a leading role in this process: the SOFICA were conceived as responding to a problem facing the sector as a whole, and intended, if anything, to favour the independents. A more congruent 'fit' between likely beneficiaries and the structure of the policy process will, on the other hand, characterize some of the remaining case studies in this chapter.

6.3 FILM PRODUCTION AND TELEVISION: SHIFTING THE BURDEN

Throughout the 1980s, the major French television stations progressively and rapidly increased their stake in the funding of film production. Thus, the 'audiovisual' media in the widest sense accounted for around ten per cent of the total investment in film production in 1985, a third in 1987, and a sum equivalent to 45 per cent of the total cost of French film production in 1990 (Farchy, 1992, p. 252). The flow of funds (that is, investment and purchase of rights) from television companies into film production grew by around 500 per cent between 1984 and 1990; moreover, in 1989 almost half of all new French films were co-produced by a television station (Farchy, 1992, p. 230).

This extraordinary shift in the funding of film production owes much to the statutory constraints progressively imposed on both private and public television stations, which include both quotas and investment obligations. In this section, we will consider only two of these measures – a law passed in 1986, which shifted the burden of funding for the *compte de soutien* to the television industry, and a 1989 law, complemented in 1990 by a decree, which imposed an obligation on all stations to devote a fixed proportion of their turnover to film production. These examples are far from exhaustive: the severe obligations imposed on the pay-TV station Canal Plus, the national quotas imposed on televisual broadcasting, or the investment obligations concerning audiovisual production had at least as great an impact on the sector. However, the measures considered here can be seen as representative of a certain policy-making pattern in the film policy arena.

The 1986 Law

In 1986, as mentioned earlier, the Socialist government oversaw the passing of legislation which reformed the funding of the *compte de soutien*. Rather than the *compte* being funded exclusively by the TSA, the financial burden was shifted overwhelmingly onto the television industry, which was subjected to a new tax representing, initially, around 4.5 per cent of the turnover of each station, the levy subsequently being increased to 5.5 per cent. The most striking characteristic of the process which led to the imposition of this tax is, as in the case of the SOFICA, the extraordinarily close co-operation between producers, civil servants and *cabinet* members in a policy network transcending the divide between the private and public sectors. Again, those involved in the process are agreed that a small number of innovating professionals set off the internal debate which led the sector to devise permanent structures for drawing the new media into the funding system centred on the *compte de soutien*. The leading figures, in this case, were the independent producer-distributor-exhibitor Marin Karmitz, who was among the professionals most frequently consulted by the Minister of Culture in the early 1980s,[22] and the head of cinema at UGC, Alain Sussfeld. However, the main producers' *syndicat*, the *Chambre Syndicale*, was, in this instance, 'very quickly involved in thinking about the issue'.[23] Again, the proposals that emerged in discussions between these professionals and the administration were taken up and articulated at the political level by Lang's adviser, Frédérique Bredin. In this case, given that television was not in the remit of the Ministry of Culture, the Ministry of Communications had to be won over – as did the Ministry of Finance, which stood to lose some of its tax revenue. With regard to the former, both the intervention of Lang and direct lobbying by the industry (including meetings between the Minister of Communications, Georges Fillioud, and the Secretary-General of the *Chambre Syndicale*, Pascal Rogard) worked to ensure the Minister's neutrality on the issue. With regard to the latter, Bredin again helped, if not to persuade the Ministry of Finance to drop its opposition, at least to ensure its technical co-operation – in the words of one *cabinet* source, 'politically, they were hostile, but afterwards, when it came to the technical analyses, the networks functioned extremely well'.[24] The success of the measure was greatly helped by the relative political weakness, at the time, of its ultimate 'victims' – the television stations. The public networks lacked any real political autonomy. Moreover, the measure was introduced just before the privatization of TF1 – as those involved in the process recognize, it would have proved impossible to impose the same concessions on the powerful private interests which owned the station from 1987 onwards. With regard to the smaller private stations which were in the process of being created, the private investors involved similarly lacked the political clout to affect the contraints being imposed. In the words of one of Lang's advisers,

> the investors at the time didn't realize quite what the percentages meant, it seemed like a minor levy basically just like any other. ... And in reality, it was very substantial. And there, we were fortunate to have an excellent understanding of the technical issues, to know and understand perfectly what was at stake, and to be dealing with people who were perhaps a little bit weaker on these points.[25]

The 1989 Law and 1990 Decree

The measures regulating the flow of funds from the broadcasting industry into the film sector were decisively reinforced in 1989 and 1990, when the Ministry of Culture initiated legislation (the law of 17 January 1989 and the implementing decree of 17 January 1990) which obliged all terrestrial television networks – public and private – to invest at least 3 per cent of their turnover in domestic film production or co-production. The process that led to the adoption of this measure featured virtually the same actors as the one described with regard to the 1986 law, with minor variations. Again, the profession and the administration shared a common objective: to standardize and render permanent the obligations which were already contained in each individual station's regulatory guidelines, the *cahier des charges*, which had until then been subject to annual renegotiation. However, the measure has to be seen against the background of the TF1 privatization, in 1986–7, when representatives of the profession were exerting strong pressure on the two main contenders for the station, Bouygues and Hachette, in order to obtain the greatest possible commitments for TF1's future investment in film production. The *Chambre Syndicale*, without openly favouring either of the two, was in close contact with Bouygues, and the figure finally written into TF1's initial *cahier des charges* – corresponding to around 3 per cent of TF1's turnover – was in large part based on an unofficial commitment made at this stage by Bouygues. This in turn served as the basis for the figure of 3 per cent subsequently chosen for all the stations concerned by the law. In the words of an industry source closely involved in both the TF1 negotiations and the preparation of the 1989 law,

> the change of mechanism [in 1989] is thus linked, in fact, to an initiative undertaken from within the profession, which consisted in obtaining this commitment from TF1, thus ensuring a steady progression of the level of investment in the film industry, as a function of overall turnover. That was the new idea.[26]

The main points of the above account can be summarized as follows. In each of the two cases related, the initiative for an adjustment in the financial relationship between cinema and television was initially articulated by influential representatives of the industry. By the time of the 1989 law, the independents who had exerted a strong influence over Lang's cinema policy in the early 1980s had to a certain extent lost their status as privileged interlocutors of the Minister and the CNC; the *syndicats* - and notably the *Chambre Syndicale* – had, by contrast, consolidated and extended their influence. It is worth noting that the role of the professional organizations – and notably the *Chambre Syndicale* – evolved considerably between the 1986 law and the law of 1989. In the words of one senior official,

> the *Chambre Syndicale* became more powerful only a long time after [1986]. At the time, the *Chambre Syndicale* was seen as somewhat outdated, and the real professionals were outside. And following this movement of 1985–86, where all of a sudden a lot of intensive consultation took place, the professionals of the film industry realized that it was in their interest to defend themselves collectively, to engage in an active partnership, instead of being unco-ordinated. ... [Before that] we met the more active producers outside the *Chambre Syndicale*.[27]

Of course, the political executive played a crucial role, in both cases, in ensuring the adoption of measures which had important implications for the government's overall audiovisual policy. However, the underlying motif in both cases is one of shared objectives, in the face of an external threat, across the boundary between the public and private sectors, and of close co-operation between officials and professionals in devising means of attaining these objectives. Thus, instead of describing the policy process in terms of a clear divide between the state and the industry, it may be helpful to consider the policy outputs discussed here as the fruit of a strong and internally cohesive policy network able to impose its aims on other actors in the politico-administrative system.

6.4 THE 'PLAN LANG' AND SUPPORT FOR BIG BUDGET FILMS

Presented as an attempt to respond to the continuing decline in audience numbers and domestic film production, the 'plan Lang' for the cinema, unveiled by Jack Lang at a press conference on 7 February 1989, comprised a number of discrete measures aimed at both production and distribution. The implementation of the 'plan' required the release of substantial budgetary credits in addition to the funds available *via* the *compte de soutien*. Thus, the Ministry's direct financial effort rose from 93 million francs in 1988 to 218 million francs in 1989. In this section, we will consider the policy process with regard to the two most important production-related measures that were announced as part of the 'plan Lang'. Unlike the policy initiatives we have considered in the previous two sections, the issues they raise relate not only to the relationship between the industry and the political executive, but also to conflicts of interest, and the distribution of power resources, within the industry. An account of the policy process should thus enable us to refine some of the observations already made about the internal structure of policy networks in the film sector.

The first measure in question was a partial reform of the *soutien automatique* – the support automatically disbursed to producers, which is proportional to a film's box office revenue. Until 1989, under a degressive scale for the reimbursement of the TSA – the levy collected on a particular film's box office receipts at its showing in cinemas – the most successful productions were 'penalized', as they qualified only for a partial reimbursement of the TSA: films with box office takings between 18 million francs and 50 million francs were eligible for a reimbursement of 85 per cent, productions with takings of more than 50 million francs received 40 per cent, while all other films received 120 per cent. From 1989 onwards, a unified rate of 120 per cent of the levy collected was applied to all productions. This had the effect of encouraging the trend towards big budget ventures, as opposed to the medium budget productions which, while accounting for the majority of productions, had most difficulty in maintaining an audience in the face of growing competition from television. The second measure we are considering here was the creation of a public *fonds de garantie*, a 'surety' fund aimed at encouraging the production, each year, of a small number of 'films à budgets importants', defined as having a budget of more than 50 million francs. The Ministry would encourage the collection of private capital for investment in these films; the public guarantee fund, set initially at 30 million francs, would cover between 50 per cent and 70 per cent of any losses the films selected might incur. This part of the 'plan Lang' is generally acknowledged by the industry to have failed. The first films selected – by a committee of producers who were themselves eligible for support from the fund – were commercial failures, and in the words of one leading producer, 'everyone split up the funds amongst themselves, and after that it was disastrous'.[28]

The 'Unification' of the *Soutien Automatique*

The reform of the *soutien automatique* had long been one of the demands of the *Chambre Syndicale*. The unification of TSA reimbursement at 120 per cent – an ostensibly 'technical' measure with important financial implications for the beneficiaries of *soutien automatique* – had been consistently demanded both publicly and in meetings with the Minister of Culture and senior CNC officials by the organization's administrator, Pascal Rogard. The pressure exerted by the *Chambre Syndicale* coincided, in the late 1980s, with a growing tendency on the part of both the administration – that is, the CNC – and the political executive to see the future of the French cinema as tied to big budget, export-oriented 'blockbusters'. Thus, in the words of a source close to the *Chambre Syndicale*,

> it wasn't very difficult to persuade Lang that one couldn't on the one hand encourage the production of big budget films ... because they were the only ones capable of helping France to withstand American competition, and on the other hand penalize the same films in budgetary terms.[29]

Yet while the general 'climate' was thus favourable to an adjustment of the *soutien automatique*, the positions adopted by the different producers' organizations were of crucial importance in conditioning the Minister's decision. Moreover, the measure enables us to make general inferences about the structure of the network concerned with film policy in the late 1980s. In particular, it provides a good illustration of the role played by the *syndicats*, and the way in which they articulated the interests of their most important members. As mentioned in the introduction to this chapter, the *Chambre Syndicale* included among its members some of the largest and most successful production concerns in the industry, notably Nicolas Seydoux's Gaumont group and the producer-director Claude Berri. In the words of a senior CNC official,

> when Nicolas Seydoux goes and argues in public – which, incidentally, he never does – that the rules governing the allocation of the *soutien* need to be changed, everyone in the industry says, 'It's outrageous, Gaumont is already super-rich, but they want even more!'

> But when it's the *Chambre Syndicale* that says, 'The rates for calculating reimbursement need to be revised', in that case, it turns into a debate within the profession and with the administration.[30]

In fact, support for the measure was far from unanimous within the industry. In very general terms, the unification of the reimbursement rate was strongly supported by the most economically dynamic and successful companies and producers. On the other hand, it was seen by less commercially oriented independent producers as a threat to the existing economic balance in the industry. According to Lang, the Ministry was concerned, in this as with other measures, to maintain a balance between the two elements of the profession:

> There was an internal debate between these two points of view. Therefore ... I tried to accomplish a kind of semantic feat, so that our support for one category did not look like a withdrawal of support for the others – one really had to find the right balance. I did not say 'supporting' big budget films, I said 'not penalizing' big budget films. 'Not penalizing' big budget films did not give the impression that we were supporting them. By saying 'supporting' experimental films we gave the impression that we were supporting them. So it was aimed at the two points of view, it allowed us to avoid a direct affront to the independent experimental directors, the independent producers.[31]

The *Union des Producteurs de Films* (UPF), with a less homogeneous membership than the *Chambre Syndicale*, was for a long time internally divided on the issue. However, once the *Chambre Syndicale* had begun to conduct a more offensive campaign on the issue, on Jack Lang's return to the Ministry of Culture in 1988, the UPF, too, saw itself obliged to adopt a definite stance on the question. The majority of its members, who ultimately shared the outlook of the dominant, Gaumont-led group in the *Chambre Syndicale*, opted to support the initiative. The UPF thus ended up following, and supporting the agenda set by, the *Chambre Syndicale*. The only significant outright opposition to the measure came from the *Association Française des Producteurs de Films*, the professional organization representing smaller independent producers. However, the AFPF lacked – and had never possessed – the same status as a privileged interlocutor of the Minister and the CNC as the larger organizations. Using the terminology of policy networks, the AFPF was very much at the margins of the film production network; the 'big' *syndicats*, and particularly the *Chambre Syndicale*, were, especially during the second Lang Ministry, indispensable partners of the Ministry at the heart of a cohesive policy community.

The '*Aide aux Films à Budgets Importants*'

The creation of a *fonds de garantie* aimed at encouraging big budget productions followed a similar pattern to the revision of the TSA reimbursement rates. Again, the initiative originated within the industry. In this instance, the idea was initially proposed to Jack Lang – and propagated among other members of the profession – by the producer Daniel Toscan du Plantier, Chairman of the Erato company and former Director-General of Gaumont. Toscan du Plantier was a leading member of the *Union des Producteurs de Films*, which threw its full weight behind Toscan's proposal for a *fonds de garantie*. The distribution of roles between the UPF and the *Chambre Syndicale* – which ultimately represented a similar range of interests – was effectively reversed when compared with the previous case study. The UPF became the project's main proponent, while the *Chambre Syndicale* adopted an at most passive stance. The two most influential members of the *Chambre Syndicale*, Claude Berri and Gaumont, were sceptical as to the eventual effectiveness of the system; Gaumont, in fact, decided not to participate in the scheme once it had been established.

However, Jack Lang, again, was easily persuaded, as the professional, sectoral logic that underpinned the project seemed congruent with the universally accepted 'cultural' aims of helping French films compete internationally and reversing the negative trend in audience figures in French cinemas. As in the case of the unification of the TSA reimbursement rates, the measure was then included in the 'plan Lang', which was publicly presented as an emanation of the political will to stem the decline of French film. On the basis of the observations we have made about the policy process with regard to the unification of the *soutien automatique* and the specific support for big budget films, the 'plan Lang' could equally be interpreted as an emanation of the dominant interests within the French film industry – though the two were not, of course, mutually exclusive.

CONCLUSION

Three central components of the policy pursued by the Ministry of Culture towards film production in the 1980s were considered in this chapter. A clear pattern of policy-making seems to emerge from these case studies. Moreover, in important respects, this pattern resembles the modes of interaction between private and public actors that were analysed in previous chapters, notably in the case of theatre policy. Again, the professional groups that benefited from public support – or more important, from state regulation – were highly organized and able to mobilize very effectively in defence of their collective interests. Again, this imposed severe constraints on the options available to the administration and the political executive. In fact, the role of the Ministry's sectoral 'clients' at all stages of the policy process went much further than in any other sector. The initiative for specific policies generally came from within the industry. In some instances, the distinction between public sector 'policy-makers' and those outside the formal structures of the state came to seem virtually meaningless with regard to the process of policy formation. It may be useful in this context to emphasize the extraordinary congruence of aims between the industry and its public backers – a phenomenon which may be seen either as an explanation or as a concomitant of this blurring of boundaries between the public and private sectors. We might also recall, at this stage, the remarks made at the beginning of this chapter about the ambiguous status of the *compte de soutien*: given the weakness, or indeed absence, of policy 'hierarchies' in the sector, it is ultimately a moot point whether the *compte de soutien* is controlled by the industry or by the state.

At a theoretical level, the centrality of the consensus between the Ministry and its 'clients' about fundamental policy objectives is captured most appropriately by the policy network approach, and more specifically, by the policy community concept as defined by Rhodes and Marsh. This is not to say that the system described in this chapter was free from internal tensions. The film production network showed a remarkable degree of cohesion when the interests of the industry as a whole were at stake. At the same time, power resources within the sector were distributed unequally, as we have tried to show in analysing the introduction of a unitary TSA reimbursement rate. Indeed, as in some of the sectors discussed in previous chapters, it is possible to discern two discrete networks: the first, centred on the integrated groups and the larger independents, corresponded to the 'policy community' paradigm; the second, more marginal network, encompassing a much wider range of actors, notably the 'smaller' independent producers, exemplified many of the traits of the 'issue network' ideal type. The correspondence between the internal structure of the profession and the varying ability of different groups of professionals to influence the policy-making process bears out one of the most important precepts of the network approach – that is, the notion that an analysis of the structure of a relevant network will contribute significantly to the explanation of a particular policy output.

In summation, we have offered two distinct, if related, analyses of policy-making in the film production sector. The first focuses on the ability of the sector as a whole to obtain satisfactory policy outcomes; the second points to the internal structure of the film policy network as a central variable in determining the shape of these policies. Both tend, of course, to undermine a statist conception of the cultural policy process, and contribute further to the emerging picture of a highly 'professionalized' cultural policy style.

7 Public Funding of Popular Music

The extension of state funding to art forms previously considered undeserving of subsidy has frequently been characterized as the most original and innovative accomplishment of Jack Lang's tenure as Minister of Culture[1]. From 1981 onwards, the Minister's public statements suggested an unprecedented willingness to accept 'popular' culture as equivalent to an established canon of 'high' culture: 'Everyone can make their choice: academic art or popular art, contemporary art or traditional art,' Lang thus declared in November 1981.[2] The new doctrine – variously denounced by Lang's detractors as heralding a dangerous cultural 'relativism' or as designed to consolidate the Minister's opinion poll ratings – was quickly translated into new funding schemes aimed explicitly at previously 'excluded' forms of expression, both 'traditional,' as in the case of 'circus arts' or *chanson*, and more contemporary, as in the case of rock music or strip cartoons.[3]

In order to compare the policy-making patterns which prevailed in the new areas of intervention with policy styles found in more traditional areas of public funding, this chapter will consider the elaboration of policy towards popular music, perhaps the most visible and controversial of the sectors to which the Ministry of Culture extended official recognition after 1981. While we will initially be concerned with gauging the role played by different actors within the core executive, the relationship between policy-makers and their 'clients' will, as in previous chapters, constitute a second focus of our analysis. The patterns of interaction between the two, however, will be found to differ radically from those observed in other sectors. Where previous case studies found stable policy communities, often based on an effective balance of power between public actors and supported institutions and individuals, policy towards popular music was based quite unambiguously on 'voluntaristic' initiatives by public officials, often using societal groups as conduits for their policy objectives and as intermediaries with target audience groups. Indeed, the present case study can be seen as a rare example, in the cultural policy area, of a sector conforming to what numerous writers on interest intermediation in France have described as a 'typically' French mode of relations between the state and societal interests – that is, a process by which public officials may decide to grant recognition selectively to certain groups, and in which the state-group relationship remains clearly skewed in favour of the representatives of the state or its component agencies. The domination, and instrumentalization, of groups by their public patrons will consequently be a central theme of the account developed in this chapter. The policy style which prevailed in the area of popular music will thus be seen to diverge in crucial respects from the dominant cultural policy style adumbrated in other chapters. However, we will also take account of the peculiar structural characteristics of the sector concerned, and relate these, as independent variables, to the policy-making patterns observed; it will then be possible to assess the implications of the analysis set out in this chapter for the formulation of more general hypotheses regarding the structural determinants of cultural policy-making.

The Lang Ministry's Support for Popular Music: a Brief Survey

Before we turn to a detailed analysis of the policy process, it may be useful, as in previous chapters, briefly to present the evolution of the Ministry's policy output between 1981 and 1993. The measures developed to support popular music differed both qualitatively and with regard to their budgetary significance from 'traditional' revenue-funding policies, and these differences were reflected, as we shall see, in the nature of the relationship between the Ministry and its 'clients.' While the 1980s saw the introduction of a wide range of ministerial initiatives aimed at 'popular music' – a vague and in some ways misleading term which, for our purposes, may be defined as comprising *chanson*, *variétés*,[4] rock and jazz – the official rationale of the policy pursued since 1981 did not aim at placing rock and other forms of popular music on an equal footing with traditional cultural policy sectors. Asked in 1985 about the relative parsimony of direct funding for *chanson*, Lang argued that it was the Ministry's aim to 'support *chanson* by means of a voluntaristic policy without turning it into another dependent and bureaucratized sector' (Lang, 1985b, p. 24). Similarly, in 1989, while conceding that the increase in support for rock and *variétés* projected in the Ministry's budget for the following year amounted to 'a great deal in relative terms, and little in absolute terms,' Lang suggested that this was explained 'directly by the fact that the sector is dominated, and has to be dominated, by the market economy' (Ministère de la Culture, 1989c, n.p.). It is difficult to establish precise figures for the Ministry's financial commitment, given the dispersal of measures among different segments of the administration. However, by the Ministry's own aggregate estimates, the budget for measures aimed at popular music, which totalled no more than 500,000 francs in 1981, amounted to around 20 million francs in 1985; by 1990, this had risen to around 40 million francs.

The Ministry's policy output in the area of popular music was characterized throughout by an unwillingness on the part of policy-makers to be drawn into the kind of long-term funding relationships with arts organizations that prevailed in other sectors; indeed, given the modest resources allocated to the new sectors of intervention, such an approach would never have been feasible. Instead, the Ministry set itself more modest objectives in three priority areas – the improvement of performance and rehearsal facilities; support for training and 'career development'; and measures aimed at stimulating the ability of the music industry itself to undertake activities benefiting the sector as a whole. Rather than provide a detailed account of each of the specific measures introduced by the Ministry under these headings, our analysis will focus on three stages in the evolution of the Ministry's policy towards popular music, each of which corresponds to a significant policy change or initiative. Specifically, we will consider, first, the initiation, in 1982, of public funding for *chanson* as the first stage in the elaboration of a policy towards popular music, as well as the subsequent shift in policy away from *chanson* and towards rock and *variétés*; second, the creation, in 1986, of a *Centre d'Information du Rock* (CIR); and third, the appointment, in 1989, of the CIR's President, Bruno Lion, to the position of ministerial *chargé de mission*, or special adviser, with overall responsibility for rock and *variétés*. The examples chosen are, of course, far from exhaustive. In particular, we do not take account of statutory and regulatory initiatives which had important economic repercussions for the 'cultural industries' in general, notably the copyright law of 1985, which established new authors' rights for performers and producers, and which created new mechanisms for distributing royalty payments within the music industry. However, issues of copyright legislation and regulation, to the extent that they concern the remuneration of individual artists and companies, are only of peripheral relevance to the subject of this book. The measures we have chosen to focus on, which were more directly linked to the funding of cultural production, will thus provide a more appropriate basis for comparison with policy-making patterns in other sectors.

7.1 FROM *CHANSON* TO ROCK: INVENTING A CULTURAL POLICY, 1981–83

The decision to extend funding to previously 'excluded' artforms followed, as we saw, from the general principle of cultural 'openness' enunciated by Jack Lang on taking office in the summer of 1981. However, the precise shape this new policy would assume was settled only after a lengthy process of experimentation. Indeed, the emphasis of policy towards popular music evolved considerably in the years after 1981: an initial focus on traditional French *chanson* thus gave way, in 1983 and 1984, to a much more visible and effective concentration of resources on rock and *variétés*, which culminated, in 1989, in the appointment of Bruno Lion as ministerial *chargé de mission* for rock and *variétés*. Tracing the decision-making processes which led first to a commitment to *chanson* and then to a dramatic shift in the Ministry's priorities towards rock and *variétés* will enable us, in this initial section, to identify the kinds of relationship that obtained both between different actors inside the Ministry, and between the Ministry and its 'clients,' and that would condition the subsequent evolution of the Ministry's policy.

The '*Politique de la Chanson*': the Definition of a 'Militant' Music Policy

From late 1981 onwards, the notion of 'cultural equivalence' articulated by Jack Lang began to be translated into concrete policy proposals. A crucial role was played here by Lang's new *directeur de la Musique et de la Danse*, Maurice Fleuret, whose personal policy priorities were similar to, if not entirely congruent with, Lang's aim of extending his Ministry's remit to previously excluded groups of cultural producers and consumers. According to a member of Lang's *cabinet*, 'one could not say that it was Maurice Fleuret more than Jack Lang, or Jack Lang more than Maurice Fleuret. It really was both of them – though each one of them had his own distinctive vision of things'.[5] Fleuret's belief that public support for music should not, as in the past, be restricted to a combination of the classical repertoire and a few 'academic' contemporaries, and his desire to, in the words of one of his collaborators at the *direction de la Musique* (DMD), 'break up all kinds of monopolies, forms of discrimination, and rackets that central and local government funding were being used for'[6] coincided with a more narrowly political interest on the part of Jack Lang and his *cabinet*. Thus, according to a senior DMD official,

> when Lang and his team arrive in 1981, their analysis is that there is this big fringe that is excluded from access to institutional culture, and so there is the will – certainly voluntaristic, but which also has an element of looking for one's sociological base – to say, right, we've got to bring on board people who are doubtless potential left-wing voters of ours, and who feel frustrated because the only official culture is an academic culture – 19th century, classical, symphonic, operatic, Boulezian, and more marginally, choral.[7]

However, while both Lang and Fleuret propounded an 'opening' of the music funding system, important choices remained as to the musical forms or sectors to which this would apply. The notion of supporting emanations of 'Anglo-Saxon' youth culture like rock music did not, at this stage, have much currency within the Ministry. Instead, the Ministry's support for popular music initially took the form of measures aimed at traditional French *chanson*; this involved, most notably, the creation of four *Centres régionaux de la chanson* (CRC), modelled on Malraux's *maisons de la culture* and devoted exclusively to *chanson*.[8] The reasons for the choice of *chanson* as the initial target of a popular music policy were both ideological and related to the personal preferences of influential members of Lang's *cabinet*. According to a senior civil servant at the DMD,

> the choice of *chanson* resulted ... from a process of lobbying that took place around Lang, and which consisted of saying: if one is going to speak of popular music ... as a left-wing civil servant, and as a representative of the Left, it needs to be legitimate by the standards of the divine, eternal Left, and so it can only be the 'quality *chanson*,' the *chanson* associated with elaborate lyrics, etc. And so, the first thing to emerge on the part of the *directeur de cabinet* – who at the time was Jacques Sallois, who had been in the PSU [*Parti socialiste unifié*, a small left-wing party that merged with the PS in the 1970s], and who had been part of a movement that had campaigned for *chanson* – was *chanson*. And Jacques Sallois's role was decisive.[9]

By contrast, the civil servants at the *direction de la Musique*, and most notably at the *division de l'Action musicale* which had been created at the beginning of 1982 to co-ordinate the DMD's policies towards popular music, were sceptical about the 'politique de la chanson,' as another DMD official points out:

> The first thing that I wasn't keen on doing, and which we were asked to do ..., was precisely the creation of these *Centres régionaux de la chanson*. The idea had not originated in the administration, but with Jacques Sallois, to be precise, who had had this idea because he came ... from an artistic *milieu*, a *milieu* of people who had talent, who were often close to the Communist Party, and who had the right ideas, except that they were out of date.[10]

While the *Centres régionaux de la chanson* were by no means the Ministry's only initiative in the area of popular music at this early stage, the extensive involvement of the *cabinet* in decisive policy decisions was indicative of the intense interest which the sector attracted at a political level. The close involvement of the *cabinet*, and of Jack Lang personally, would similarly be evident at the next stage of policy towards popular music – which involved a dramatic switch of support away from *chanson*, and to the more 'promising' sectors of rock and *variétés*.

From *Chanson* to Rock and *Variétés*: Redefining a Cultural Policy

It quickly became apparent both to the civil servants at the *division de l'Action musicale* and to the *cabinet* members following the sector that the focus on *chanson* would never have the mass public impact that was at least one of the intended consequences of the 'opening' of the funding system to new forms of expression – not least because traditional French *chanson* had long been eclipsed by 'Anglo-Saxon' imports such as rock as a dominant genre of popular music. Moreover, the institutional form chosen for the *chanson* policy, the *Centres régionaux de la chanson*, was widely seen as a failure even after the first year. In the words of one DMD official, 'it would have been suicidal to continue, all of them were in deficit and in crisis'.[11] The crisis of the CRC was magnified by the tenuous funding framework initially established by the Ministry. The CRC had been conceived on the assumption that their funding would be taken over in large part by the regions, as the Ministry was unwilling or unable to permanently fund a new regional infrastructure potentially subject to the same problems of institutional *lourdeur* as the *maisons de la culture*. The problem is summarized by a former DMD official: 'There was also a kind of transfer ... which involved saying: "I, the state, am taking the decision, but the ones who are going to have to foot the bill are the regions, or the municipalities." And the regions very quickly understood that this was all very well, but that it should rather be central government that pays.'[12] Against this background, the Ministry decided to withdraw its funding progressively from the *Centres régionaux de la chanson*. According to another DMD official,

> very quickly, we realized that there was a kind of misunderstanding, a bad start – you can even call it a very bad start, the wrong kind of start – for this policy, which could have been very serious. And so all these *Centres régionaux de la chanson* existed ..., and we quickly realized that it wouldn't work. ... And so what happened is that, throughout 1983 and 1984, we started to liquidate what had been put into place in 1981 – which makes for a strange beginning.[13]

The decision to redirect the funds initially allocated to *chanson* was taken, again, at the level of the *cabinet*, in consultation with the *directeur de la Musique* – though the impulse came, not least, from the officials at the *division de l'Action musicale*, as one of the latter recalls: 'We lobbied incessantly to put a stop to it, and Maurice Fleuret was sufficiently powerful to be able to decide rapidly, and we were able to put a stop to it.'[14]

As our brief account of the Ministry's policy towards *chanson* and the *Centres régionaux de la chanson* has underlined, the nature of policy-making towards popular music diverged in important respects from the policy styles prevailing in other sectors. Most obviously, the structural framework of the Ministry's intervention was elaborated in close collaboration between the *cabinet* and the sectoral bureaucracy, and thus subject to extensive and direct political control. Second, and perhaps equally important, the professional *milieu* at which the *chanson* policy was primarily directed was politically and economically marginal and lacked any degree of autonomy *vis-à-vis* the Ministry. Third, and following on from the previous points, the specific policies elaborated were subject to rapid and thorough revision on terms defined entirely by the Ministry. In subsequent phases of the Ministry's popular music policy, these traits were to remain essentially unchanged; while the role of selected private-sector actors in the elaboration and implementation of policy assumed progressively greater importance, the policy process retained its hierarchical and 'politicized' character.

7.2. THE INVENTION OF A 'ROCK POLICY': THE *RÉSEAU ROCKS* AND THE CREATION OF THE *CENTRE D'INFORMATION DU ROCK*

In parallel with the termination, or at any rate relegation, of the ill-fated *politique de la chanson*, the Ministry began to target more relevant and visible forms of popular music, notably *variétés*, jazz and, perhaps most important in view of the demographic composition of its primary audience, rock music. According to a former DMD official, the 'disenchantment' with the policy towards *chanson* coincided with the realization on the part of the administration that 'if one really was going to engage in this kind of voluntaristic work of trying to integrate a category of the population that was excluded from access to culture, then it had to be based on rock and jazz'.[15] As in the *chanson* 'phase' of policy towards popular music, the acuity of political interest in the emergence of a 'rock policy' was reflected in the active and consistent involvement of Jack Lang's *cabinet* in the elaboration of the Ministry's stance on rock and *variétés*, and, less directly, in the personal interest which the Minister himself showed for the projects initiated by the DMD. In general, the role of the *cabinet* was limited to supporting the initiatives proposed by the civil servants of the *division de l'Action musicale*, not least against the more sceptical attitude of other, traditional sectors within the music administration. As one DMD official recalls,

> we would get remarks from the artistic inspector at the *direction de la Musique*, who told us, 'Really, we need the money in other areas, is it reasonable to be spending 20 or 25 million francs on jazz or on *chanson*, perhaps there are more important things to do.' And so in cases like that, we relied on the support of the *cabinet* to point out that the Minister was interested in this, that it was also important from an image point of view.[16]

However, while the general impulse to redirect the focus of the Ministry's efforts towards rock, jazz and *variétés* originated in the *direction de la Musique*, some members of Lang's *cabinet* quickly became involved in particular elements of 'rock policy.' Specifically, Lang's architectural adviser, Christian Dupavillon, who had played a crucial role with regard to the initial choices defining the *grands travaux* programme, was to oversee the construction of the *Zénith*, a 'temporary' hall, designed especially for pop concerts, which was opened in January 1984 at the Parc de la Villette in Paris.[17] More generally, according to a senior DMD official, 'with regard to rock, we soon had a very strong supporter in the *cabinet*. This was Dupavillon, who, incidentally, quickly overtook us, in the sense that, in virtue of his relationship with Lang and his personal energy, he became more effective than what we could put into place at the *direction de la Musique*.'[18] The exceptional degree of political 'voluntarism' evident in the role played by the Minister's *cabinet* was not, however, sufficient in itself for the successful development of a policy on popular music. Above all, the Ministry lacked credible private-sector or professional partners in the rock and *variétés* sector; much of the Ministry's initial work, once the decision to devote resources to rock and *variétés* had been taken, was thus devoted to identifying suitable groups or institutions at whom the policy might be aimed.

'Rock Policy' and its Conduits: the 'Co-optation' of the *Réseau Rocks*

It was clear from the outset that a sector like rock or *variétés*, which had until then been operating exclusively along commercial lines, would not be amenable to the forms of intervention that had prevailed in traditional performing arts sectors. The Ministry thus sought to establish links both with the music industry and with voluntary groups active in the amateur rock *milieu*. In the case of *variétés*, the development of a coherent policy was facilitated by the existence of a well-organized and economically powerful industry, which could be integrated into the elaboration of specific measures by means of a network of representative organizations and consultative committees. The Ministry was thus able very quickly to establish a small number of institutions and programmes aimed specifically at improving the training of *variétés* artists, which were co-funded by the music industry *via* the various statutory authors' and performers' rights organizations. In the case of rock, a sector potentially of much greater political importance, establishing public support schemes and finding appropriate professional interlocutors proved more difficult, not least because many of the potential 'clients' of a 'rock policy' viewed the notion of state intervention in the sector with suspicion. Even before 1981, a small number of municipalities had begun to grant modest subsidies to local voluntary groups concerned with improving concert and rehearsal facilities; however, no such contacts had existed at the national level. Nonetheless, a potential partner for the Ministry's policies emerged in the form of the *Réseau Rocks*, a loose federation of local *associations* representing amateur or semi-professional rock groups. In the absence of obvious institutional or commercial conduits for a national 'rock policy,' the *Réseau Rocks*, without possessing a strong organizational structure or being in any formal way 'representative' of the amateur rock *milieu*, quickly established itself as the Ministry's privileged interlocutor for all issues relating to rock.

The *Réseau Rocks* was by no means the only national network of *associations* concerned with rock issues, and its status as the Ministry's chosen partner in the sector owed much to purely contingent factors. However, two features set the *Réseau Rocks* apart from other rock *associations*. The first was the importance its members attached to obtaining official 'recognition' and subsidies for rock, which contrasted with the more sceptical and at times hostile attitude of some other groups. The second was the personal access of the *Réseau*'s President, Bruno Lion, to policy-makers, initially in the Ministry's regional directorate for Ile-de-France, and later in the Ministry itself. As one of the founders of the *Réseau* recalls,

> just as each one of us in the provinces was talking to the representatives of local government, Bruno [Lion] was talking to people at the national level – in fact, he had a link to the DRAC [*direction régionale des Affaires culturelles*], and then very rapidly to the Ministry of Culture.[19]

The 'legitimation' of the *Réseau* as the Ministry's main interlocutor was facilitated, indirectly, by the fact that Lion's father, Robert Lion, occupied a succession of senior administrative posts under the Socialist governments of 1981–6 – as *directeur de cabinet* to the first Socialist Prime Minister, Pierre Mauroy; as a member, alongside Lang, of the 'Group of Four' which was initially given the task of piloting Mitterrand's *grands travaux*; and subsequently as director-general of the parastatal *Caisse des dépôts et consignations*.

From the outset, Lion and the *Réseau Rocks* were closely associated with both the elaboration and the implementation of the Ministry's policy on rock. Thus, the publication, in December 1984, of a 'guide' to rock rehearsal venues, entitled 'Maxi-Rock, Mini-Bruits,' resulted directly from a meeting held in July 1983 between Lion and officials at the Ministry's *direction du Développement culturel* (which, in parallel with the *direction de la Musique*, was beginning to take an interest in rock, though less from an artistic than from a 'socio-cultural' perspective). Similarly, the creation, in 1985, of 200 temporary posts in voluntary rock *associations*, financed by the Ministry of Culture and intended to develop the sector's internal organization and economic potential, was a direct response to a demand made publicly in October 1984 by Lion.[20]

Towards the end of the 1981–6 legislature, the relationship between the Ministry and the *Réseau Rocks*, which had initially been based on collaboration on specific projects, assumed a more permanent and institutional character. In September 1985, Lion and another founder of the *Réseau Rocks*, Maurice Lidou, met the civil servants responsible for 'rock policy' at the *direction de la Musique* and suggested a range of more permanent support schemes which would deepen the Ministry's commitment to the sector. Of these proposals, the DMD decided to support an initiative for the creation of a permanent, publicly funded rock 'information centre,' to be established and administered by the *Réseau Rocks*. The objectives of the new institution – described officially as an 'inter-regional information network intended for rock musicians and for the various partners concerned by their activities'[21] – were ostensibly modest in their scope; however, the 'information centre' would later acquire a central role, beyond the mere provision of 'information,' as something resembling an implementing agency for the Ministry's 'rock policy.'

In fact, the DMD's decision, in the autumn of 1985, to opt for the creation of an 'information network' as a centrepiece of its 'rock policy,' and not to embark on more direct forms of support for rock, was largely motivated by political considerations relating to the imminent 1986 legislative elections. The formal convention between the Minister of Culture, the Minister for Youth and Sports, and Bruno Lion as President of the *Réseau Rocks*, establishing the 'information network' that was to become the *Centre d'Information du Rock*, was thus signed only on 6 March 1986, a few days before the elections which initiated the first *cohabitation*. According to one of the founders of the *Réseau Rocks*, the formula of an 'information centre' was chosen 'for entirely pragmatic reasons – that is, it was politically more neutral, and with the right-wing government that was going to come to power six months later, it seemed perhaps the best bet to get started on'.[22] Notwithstanding its initially limited competences, the creation of the CIR signified a new stage in the evolution of a national 'rock policy.' Without committing itself to the direct funding of individual artists, the Ministry established a long-term, and effectively irreversible, funding relationship with a non-governmental institution.[23] Conversely, the *Réseau Rocks* ceased to be a societal 'pressure' group and mutated into a 'client' group whose legitimacy depended entirely on its accreditation and funding by the Ministry. Indeed, the *Réseau Rocks* was dissolved a year after the creation of the CIR: in the words of one of its founders,

> the most active people in the *Réseau Rocks* get involved in this project, which means that from the moment where the CIR is created, the entire energy of the *Réseau Rocks* is mobilized for it. And as a result, the *Réseau Rocks* is completely de-activated.[24]

The creation of the CIR as a semi-official conduit for the measures elaborated by the DMD, and the concomitant dissolution of the *Réseau Rocks*, signified the accomplishment of the first phase of the Lang Ministry's 'rock policy.' Two central features of the policy-making process in this period should be underlined. The first concerns the distribution of competences between the sectoral administration and Jack Lang's *cabinet*. Thus, while the initiative for the Ministry's first, tentative experiments in the funding of rock and *variétés* projects can be traced to a small group of civil servants at the *direction de la Musique*, the development of a 'rock policy' was closely watched, and supported, by the Minister's *cabinet*; in some instances, individual *cabinet* members played a still more active role with regard to particular projects. The second element that should be emphasized in our account of policy-making before 1986 concerns the relationship between the Ministry – in this instance, the *direction de la Musique* and, by implication, its political superiors – and societal groups affected by its policies. When the Ministry began experimenting with funding policies for rock music, no obvious professional or institutional partners existed to absorb these initiatives, not least as a consequence of the lack of clear distinctions between professional and amateur groups in the rock *milieu*.[25] The Ministry's relationship with the *Réseau Rocks*, which quickly emerged as a suitable conduit without being either heavily institutionalized or in any strict sense 'representative' of the *milieu* as a whole, consequently diverged in crucial respects from the patterns we observed in other sectors. As we have already suggested, the notion of 'co-optation' and deliberate legitimation of groups by the state, often described as a paradigmatic mode of state-group relations in France, may offer a more appropriate framework for conceptualizing the role of extra-governmental actors in this area than the policy network approach. The issues raised here will be discussed at greater length in a later section of this chapter; first, however, we will consider the third and final phase of the Lang Ministry's 'rock policy,' which correponds to the period from Bruno Lion's appointment as ministerial adviser for rock and *variétés* in 1989 to the legislative elections of March 1993.

7.3 1989–93: BRUNO LION AND THE NORMALIZATION OF 'ROCK POLICY'

As the outgoing Minister of Culture and his administration had hoped, the institutions and programmes created in the rock sector just before the victory of the Centre-Right in the 1986 elections survived the *cohabitation* period unscathed. Thus, while neither François Léotard, the new Minister of Culture, nor his *directeur de la Musique*, Marc Bleuse, had much personal enthusiasm for the notion of state support for rock and *variétés*, the political cost of abolishing an institution like the *Centre d'Information du Rock* appeared prohibitive; the popularity of Lang's decision to extend subsidies to artforms beyond the established canon had by now been firmly established – in February 1985, an opinion poll thus found 77 per cent of respondents favouring the Ministry's support for popular music.[26] Consequently, while budget lines for existing programmes were frozen or regressed slightly, and no new initiatives were launched during *cohabitation*, as in most other sectors, François Léotard's tenure as Minister of Culture left no significant traces on the policy pursued by the Ministry *vis-à-vis* rock and *variétés*.

On Lang's reappointment as Minister of Culture in 1988, the process initiated towards the end of the first Lang Ministry – the 'institutionalization' of state support for rock and *variétés* – could thus be continued. The central features of the policy-making process identified in our account of the 1981–6 period remain applicable to this later phase in the Ministry's 'rock policy.' However, the distribution of roles and influence among the main actors concerned with rock and *variétés* after 1988 differed in important respects from the pattern identified with regard to the earlier period. Given the media attention which the funding of popular music had generated before 1986, and the proven popularity of the measures, Lang and his *cabinet* had by now become convinced of the benefits of pursuing the Ministry's interventions in the rock and *variétés* sector in a more direct and visible fashion. In June 1988, Lang consequently asked Lion, President of the CIR and former President of the *Réseau Rocks*, to assume a formal position within the Ministry, with a view to co-ordinating the entirety of its policies in the domain of popular music and, more important, developing new funding initiatives.[27] Thus, in May 1989, after a transitional period during which Lion carried out an informal 'mission de réflexion' on the Ministry's behalf, he was appointed *chargé de mission*, directly answerable to the Minister, with responsibility for rock and *variétés*. While Lion was not, formally, a member of the *cabinet* and had neither direct hierarchical authority over the civil servants of the *direction de la Musique* nor an independent budget, the appointment was intended both to raise the public profile of the Ministry's intervention in the sector and, in the words of one of Lang's advisers, to

> put some order and logic back into a policy which had been conducted in a very intuitive way, ... to give it all a spinal column, an overall rationale which had not until then existed, except in the form of something that was improvised retrospectively...[28]

Lion's appointment received extensive coverage in the media, and Lion, quickly dubbed 'Monsieur Rock' by the press, subsequently remained, in his own words, 'extremely exposed to the media'.[29] However, there were more substantive implications both for the evolution of the Ministry's policy output and with regard to the nature of the policy process. The programmes initiated before 1986 remained in place; on the other hand, the policy initiatives additionally put forward by Lion marked a departure, in important respects, from the modes of intervention which had been elaborated by the Ministry's civil servants in the early 1980s. Lion's major innovation, the *Fonds d'Action et d'Initiative Rock* (FAIR), developed in the course of the 'mission de réflexion' that preceded his appointment, and officially announced by Jack Lang in September 1989, soon after Lion's arrival at the Ministry, exemplified the difference in approach. The FAIR's principal aim was described as 'supporting the career start of artists resident in France';[30] in concrete terms, the FAIR involved the selection, by representatives of the record industry, of 15 French groups or artists each year, who were offered training and professional advice as well as limited financial support to cover touring expenses and equipment for a one-year period. During the first Lang Ministry, the *direction de la Musique* had consistently and consciously avoided the attribution of direct subsidies to individual artists – partly because of the sector's budgetary limitations, but also, according to one of the officials concerned, because 'we wanted to avoid falling into the contemporary musicians' syndrome'.[31] Lion's FAIR programme, by contrast, entailed the disbursement of direct financial assistance to particular groups or performers. Moreover, while earlier measures had been aimed largely at amateurs or semi-professionals, the FAIR clearly implied a 'professionalization' of 'rock policy.' The funding distributed by the FAIR – which did not exceed 40,000 francs for each group selected – was insignificant when compared with the institutional subsidies of 'high' culture, and the programme was co-funded by a private sponsor and by a number of musical authors' and performers' rights societies; nonetheless, it marked a break with the indirect funding patterns which had characterized 'rock policy' before 1989.

In parallel with this shift in the focus of the Ministry's funding efforts, Lion's appointment signified a change in the way in which policy was elaborated. Specifically, the creation of the FAIR programme, drafted entirely by Lion with the help of the CIR and its contacts in the professional *milieu*, offers a clear illustration of the extent to which primary responsibility for popular music had by now been taken out of the hands of the Ministry's sectoral administration. We noted that a crucial 'pioneering' role was played by 'regular' civil servants, notably at the *direction de la Musique*, in initially defining the general form of state intervention in the area of popular music, and in identifying specific target groups for such a policy. Towards the end of the first Lang Ministry, this role was increasingly supplemented, and to some extent supplanted, by the elevation of the *Réseau Rocks* to the status of official 'mediator' between the Ministry and the rock *milieu*. By 1988, after two years of *cohabitation*, the DMD had effectively relinquished its role as policy innovator.

The relative marginalization of the DMD was further accentuated by the personnel changes that had occurred since 1986. Most of the officials who had played a central role in the defining phase of the Ministry's 'rock policy' had by now moved to other segments of the administration. Perhaps most important, the *directeur de la Musique* appointed by Lang in 1988, Michel Schneider, disagreed fundamentally with the principle of state support for what he considered 'inferior' forms of expression, and saw the funding of popular music as little more than a public relations exercise.[32] While Schneider's opposition remained latent at first, it increasingly contributed to his conflicts with Lang; in any case, it prevented the DMD from playing a significant role in the definition of new policies. Schneider's 'obstructionism' risked reversing the significant gains in sympathy which the Lang Ministry had been able to make among the policy's target constituencies; indeed, according to one of Lang's advisers, the decision to place the Ministry's policy towards rock under more direct political control stemmed, not least, from the need to respond to 'pressure from the grassroots that was even stronger, even more threatening at that moment given that the then *directeur de la Musique* – that is, Michel Schneider – did not care at all about the sector'.[33]

The sectoral administration, in the form of the DMD and, to a lesser extent, the Ministry's *délégation au Développement et aux Formations* (DDF), continued to be formally responsible for funding specific measures, even where these had been elaborated by Bruno Lion. Lion's role inside the Ministry thus consisted largely in negotiating funding for particular initiatives both with the relevant *conseillers techniques* in Lang's *cabinet* and with his counterparts in the DMD and DDF. Not surprisingly, Lion's relations with Schneider were tense, and one of Lang's advisers notes that Lion 'spent a lot of time crossing swords with Schneider'[34]. However, as in other areas where Schneider found himself in disagreement with Lang's policy options, such as the contemporary music sector, where Schneider criticized the role played by Pierre Boulez in the elaboration of the musical *grands travaux*, his impact on policy was very limited. Lion was able to doubly circumvent the *directeur de la Musique*, both by dealing directly with the middle-level civil servants at the DMD with whom he had already established a close relationship in his capacity as President of the *Réseau Rocks* and CIR, and, perhaps more important, in virtue of his privileged relationship with Lang. Thus, Lion's status as a manifestly political, rather than technical, appointee, responsible for a politically 'sensitive' sector, naturally bolstered his leverage *vis-à-vis* the administration and *cabinet*; it may also account in part for the substantial expansion in the volume of funds allocated to rock and *variétés* by the Ministry in Lion's first year as *chargé de mission*.[35]

At another level, Lion's peculiar position outside, or above, the established administrative hierarchy ensured an unusual degree of ministerial involvement in the elaboration of policy. In concrete terms, this meant that important policy proposals could be submitted directly to Lang by his *chargé de mission* for rock and *variétés*: as Lion recalls,

> especially during the initial period ... it was always possible for me to go into his office and say, 'Here we are, this is what's happening, I suggest we do that, do you agree, yes or no,' and within three minutes, the answer was there, you had a clear and unambiguous answer.[36]

The explicitly 'political' nature of the policy-making process in the domain of rock and *variétés*, and its detachment from the procedures of sectoral administrations like the *direction de la Musique*, which had been evident since the earliest experiments with the funding of popular music at the beginning of the first Lang Ministry, were, in other words, significantly strengthened during the 1988–93 period.

Before we conclude our account of this final stage in the development of the Ministry's 'rock policy,' it may be worth adumbrating the evolution, in the same period, of the Ministry's relations with its 'clients' in the area of rock and *variétés*. As mentioned earlier, Bruno Lion aimed to 'professionalize' the focus of the Ministry's interventions, and, by means of the FAIR programme, introduced a form of direct, if modest, subsidy for the first time. This did not, however, imply a fundamental shift in the nature of the relationship between the Ministry and the *milieu*. While Lion's strategy for developing the Ministry's policy was based on extensive and institutionalized consultation with the music industry, notably through the creation of a 'consultative committee' for rock and *variétés*, and in the context of the selection procedure for the FAIR programme, the role of the Ministry's 'clients' in defining policy objectives remained marginal. In fact, the music industry's initial reaction to the FAIR, the most important of the initiatives developed by Lion after 1988, was at best ambiguous. According to one of Lang's advisers,

> at first, it brought [Bruno Lion] into conflict with the music industry. Because they saw it as a form of competition, as meddling in their affairs, they didn't see the possibility of a partnership. ... [The record companies] tried to block things, if anything, and they had to be won over one by one.[37]

Characteristically, then, far from lobbying for public support, the major private sector actors in the rock sector had to be actively persuaded of the benefits of the measures proposed by the Ministry. This mode of policy elaboration and implementation contrasts starkly with the pattern discerned in previous chapters of this book, where important policy initiatives in traditional areas of arts funding were often found to be based on the adoption, by the administration and political executive, of priorities defined and articulated by the Ministry's 'clients.' The continuing detachment of the music industry from the policy-making process, both in this particular case and *vis-à-vis* the Ministry's popular music policy in general, can be seen as a function of the marginal impact the Ministry's support programmes made on the economics of the sector as a whole: given the modest volume of credits devoted to 'rock policy,' and given the dispersal of funding among a large number of beneficiaries, only a very limited level of interest in the formulation of policy could be expected from the Ministry's 'clients.' Conversely, the absence of substantial private-sector inputs into the policy process further underlined the pre-eminence of a hierarchical, highly politicized policy style focussed on the Minister and his advisers.

CONCLUSION

Two closely related dimensions of the policy-making process with regard to popular music were considered in this chapter: first, the influence of different actors inside the core executive – notably the Minister, his *cabinet* and relevant segments of the bureaucracy – both on the initial decision to develop a set of policies aimed at popular music and on specific policy measures; and second, the impact of extra-governmental groups, both amateur and professional, on the Ministry's intervention in this area. With respect to both of these, we found that policy-making towards popular music occupied a unique place among the sectors considered in this book. Indeed, the account developed here suggests that, quite unlike any of the traditional performing arts sectors supported by the Ministry of Culture, policy towards popular music conformed to a hierarchical, state-centred conception of cultural policy-making, in which the entirety of the Ministry's output in the policy sector concerned can be seen as flowing directly from a decision made at the level of the political executive. Of course, the policy could not be attributed exclusively to the Minister and his *cabinet*. The general aim of extending funding to non-classical popular music was shared by Lang's first *directeur de la Musique*, Maurice Fleuret, and by some senior officials in the *direction de la Musique*. Moreover, most of the specific measures introduced under the first Lang Ministry were elaborated by civil servants in the Ministry's *direction de la Musique*, with, initially, only occasional involvement by the Minister or his *cabinet*. However, while 'regular' civil servants were responsible for developing a range of specific support mechanisms, and while the *direction de la Musique* had a significant degree of influence over the choice of funding partners and beneficiaries in the broad area of popular music, the political importance attached to the sector by the Minister and his advisers was never in doubt. The output produced by the *direction de la Musique* and other segments of the administration was closely supervised by the *cabinet*, and certain policy initiatives were handled almost exclusively by particular *cabinet* members. Moreover, the importance attached by Lang to the effective public presentation of the policy in itself ensured a high degree of conformity between the Minister's priorities and the measures developed at other levels of the ministerial hierarchy. The effective delegation of responsibility for policy on rock and *variétés* to Bruno Lion at the beginning of the second Lang Ministry served to further accentuate the extent to which ultimate control of the policy process lay at the ministerial level. In short, the account developed in this chapter points to a relatively unambiguous instance of 'ministerial government', following the definition suggested by Elgie (1993, p. 32), whereby the Minister is seen as the 'chief policy-maker' in a particular policy area.[38]

Our analysis of policy-making at the level of the core executive was complemented with an account of the relationship between the Ministry and its private sector 'clients'; here, too, we found few similarities with the policy styles prevailing in most other sectors. Most crucially, we noted the absence, in the popular music sector, of professional or industrial producer interests significantly affected by the policy outputs of the Ministry of Culture. Given the lack of plausible institutional interlocutors, the *Réseau Rocks*, a weakly organized federation of local rock *associations*, with an uncertain membership base and without clear representative credentials, acted as a conduit for the Ministry's policies. As the Ministry's 'rock policy' became increasingly institutionalized, the *Réseau Rocks* progressively lost its autonomy *vis-à-vis* its administrative and political patrons, a process which culminated in the creation of the *Centre d'Information du Rock* and, finally, in the appointment of Bruno Lion as *chargé de mission*. The Ministry's relationship with the *Réseau Rocks* thus exemplifies a specifically French mode of interest group politics described by Mény (1989, p. 361) as 'integration and co-optation by means of legitimation,' by which policy-makers selectively accord official recognition to 'tame' interest groups in order to facilitate policy implementation or gain the support of certain societal target groups. Two central features of this phenomenon should be underlined: first, the organizational weakness of the group concerned as a precondition of 'co-optation', and second, the loss of autonomy which results from 'co-optation'. Examples of this 'process of selection-integration' (Mény, ibid.) abound outside the cultural policy arena, and links with selected 'non-occupational advocacy groups' (Wilson, 1987, p. 51) formed an important component of the French Socialists' electoral strategy throughout the 1980s: the case of Harlem Désir's *SOS-Racisme* could thus be seen as emblematic of the relationship between the PS and social movements focused on 'new politics' issues. This is not to suggest that the *Réseau Rocks* should be seen as a 'new social movement' or indeed that the *Réseau Rocks* ever had a significant capacity for mobilizing even the rock sub-culture, let alone an entire cohort of electors. Indeed, the characterization of the *Réseau Rocks* as a societal 'advocacy group' is questionable: given the exceptional nature of the links between Bruno Lion and the Ministry, and Lion's personal access to key members of the Socialist policy-making elite, it is impossible to assess with any accuracy the role played by the *Réseau Rocks* as a group, independently of Lion's personal role. Nevertheless, the analogy succeeds in isolating the central characteristics of the Ministry's relationship with the *Réseau*, and in identifying a general category of state-group relations in which it might be situated.

As a concomitant of the account developed above, we might note that the policy network approach – used as a tool of analysis in most other sectoral case studies in this book – has only limited value as a framework for conceptualizing the policy-making process in the area of popular music. Given the dominant role of governmental actors both in setting the policy agenda and in devising particular measures, and given the absence of significant and autonomous extra-governmental actors, the application of the comparative framework used in other sectors would provide meagre hermeneutic benefits. However, the policy network approach is in no way diminished or invalidated by this finding. The policy studied had important symbolic repercussions, yet its economic significance was extremely limited. That we should find only a very limited degree of professional mobilization in a sector affected only at the margins by public intervention is hardly surprising; in the absence of a significant 'issue' affecting important elements of an industry or profession, we should not expect the formation of significant networks. Thus, while the potential benefits of the policy were keenly perceived at the political level, this corresponded in no way to perceptions on the part of the private-sector groups who might otherwise have played an important role in the policy-making process. In conclusion, it would perhaps be unjust to adopt the assessment of a senior Ministry official, closely involved in shaping the policy in its initial phase, who observes that 'quite frankly, I had a great and exciting time in 1981–83; by 1984–85, I had had enough, I already more or less had the feeling that we were being used as an alibi for a cultural policy that was ultimately quite rigid, quite conventional'.[39] Nevertheless, we may note that the significance of our account of the Lang Ministry's popular music policy for a more general analysis of the determinants of cultural policy lies, above all, in the contrast the sector offers with 'established' cultural policy sectors – both in respect of the form of public intervention and with regard to the prevailing policy-making style.

8 The Politics of *Développement Culturel*

The analysis of the policy process developed in previous chapters focused on the relationship between the Ministry of Culture and its 'clients' in particular sectors, and emphasized the influence of sectoral interest groups on the Ministry's policy output. In this chapter, by contrast, we will be considering a sector characterized by the virtual absence of stable funding relationships between government and private sector 'clients', and in which policy can, by definition, be driven only by a 'political' dynamic. The choice of *développement culturel*, or cultural development, is not fortuitous. The central theme of this book so far has been the 'stabilizing' impact of sectoral policy networks on policy outputs. By contrast, the dominant theme in this chapter will be the extreme vulnerability to a wide range of exogenous influences – many of them contingent and personal – of a policy area marked by the absence of strong, integrated policy networks. The account set out in this chapter will thus, in a sense, act as a controlling case for the sectoral case studies of previous chapters. Instead of looking at individual policies, this chapter will consider the factors that influenced the budgetary and institutional status of *développement culturel* within the Ministry of Culture between 1981 and 1993. First, however, we will outline what is understood by *développement culturel* and sketch the role it has played in recent French cultural policy.

8.1. CULTURAL 'DEMOCRATIZATION', 1959–93

Développement culturel does not, unlike the areas of state intervention considered in previous chapters, merely denote a set of policies developed to meet the needs and concerns of a specific sector of cultural production. The generic term *développement culturel* refers, rather, to a wide and disparate range of policies expressly aimed not at stimulating cultural production, but at what is generally, and crudely, termed the 'democratization' of culture. Officially, cultural 'democratization' – which might take the form of facilitating access to established cultural institutions, or of improving arts-related education, or of encouraging the amateur arts – has been a central, indeed the central, policy objective of the Ministry of Culture. The stated aim of the Ministry of Cultural Affairs created by André Malraux in 1959 was to 'make the great masterworks of mankind, and above all of France, accessible to the greatest possible number of Frenchmen, provide the largest possible audience for our cultural heritage, and encourage the creation of works of art and of the spirit which will enrich it' (quoted in Wangermée, 1988, p. 29). It is worth noting the ordering of priorities implied here: making the arts 'accessible' was presented as the Ministry's primary *raison d'être*. For Malraux, this meant little more than reducing geographical and other physical barriers to the consumption of a universally valid 'high' culture, not least by aiming to construct *maisons de la culture* throughout France. In the early 1970s, Malraux's doctrine gave way to an approach that allowed for a more flexible and heterogeneous interpretation of 'culture', and which acknowledged that 'democratization' might equally imply the recognition, and encouragement, of diverse forms of expression outside the established canon of 'high' culture. The term *développement culturel*, which was first used in this period, was later adopted by the Socialists to describe their own policies in favour of 'democratization'.

Little would be gained, in this context, by entering into the nuances of the ideological debates surrounding the notion of *développement culturel*. However, two general remarks should be made about the historical development of this facet of the Ministry's output. First, 'democratization' as an official policy goal has assumed great symbolic importance for policy-makers in their efforts to legitimize the existence of a Ministry of Culture, and to lend some degree of coherence to its discrete and sectorally differentiated spheres of intervention. Indeed, Urfalino (1993b) suggests that the 'democratizing' component of the Ministry's policy could be construed as the symbolic 'guarantee' that its 'cultural policy' adds up to more than merely an uncoordinated assortment of sectoral, producer-directed funding policies. Specifically, Urfalino argues that the related goals of 'democratization' and decentralization have been the single unifying rationale of the Ministry since its inception under Malraux in 1959, and that the very legitimacy of the appellation 'cultural policy' hinges crucially on the kind of intersectoral activity implicit in *développement culturel*. The second, related, point concerns the modest institutional guise in which the policy of 'democratization' has been pursued since 1959. In spite of its symbolic importance, the resources allocated to this area of the Ministry's policy have never been on a par with the funding accorded its traditional 'clients'. In 1975, only 3 per cent of the Ministry's budget was programmed specifically for measures relating to *développement culturel*; by 1981, this proportion had declined to a derisory 1.5 per cent. The subordinate status of *développement culturel* was similarly emphasized by the Ministry's organizational structure. While each 'traditional' sector was administered by a *direction*, the sector of *développement culturel* was entitled only to a less prestigious administrative structure; in 1981, this took the form of a *mission de développement culturel*, the main responsibility of which was the funding of, and administrative control over, the network of *maisons de la culture*. In addition, a small interministerial institution attached to the Ministry of Culture, the *Fonds d'intervention culturelle* (FIC) was engaged in supporting individual 'innovative' projects aimed explicitly at 'new' or 'potential' arts audiences. It should be added, however, that between 1971 and 1981, its average annual budget amounted to no more than 14 million francs. Both the budgetary situation of *développement culturel* and its institutional status within the Ministry underwent significant change with the advent of a Socialist government in 1981. Yet as the next section will suggest, after an initial, politically-motivated 'boost' in its budget and status, *développement culturel* returned more or less to the relative institutional stagnation which had characterized it before 1981.

Développement Culturel: a Linchpin of Socialist Cultural Policy?

In discussing the policy of *développement culturel* pursued during Lang's tenure as Minister of Culture, it is important to distinguish between an initial period of budgetary and institutional expansion, which lasted only until 1984, and a later period in which *développement culturel* was once again relegated to a subordinate position among the Ministry's priorities. The first, expansionary period was dramatically initiated by Lang's first budget as Minister of Culture: from 41 million francs in 1981, the budgetary allocation of *développement culturel* rose to 710 million francs in the 1982 budget, almost the equivalent of the budget of a 'traditional' sector like theatre or music (see Figure 8.1).

The budgetary expansion coincided with the creation of a separate *direction du Développement culturel* (DDC), which took effect in May 1982 and similarly signified an 'elevation' to the level of traditional funding sectors. The responsibilities of the DDC were potentially vast. The role of the new *direction* was to be 'transverse'; rather than restricting itself to specific projects related to 'democratization', the DDC would have the task of 'horizontal' co-ordination among the

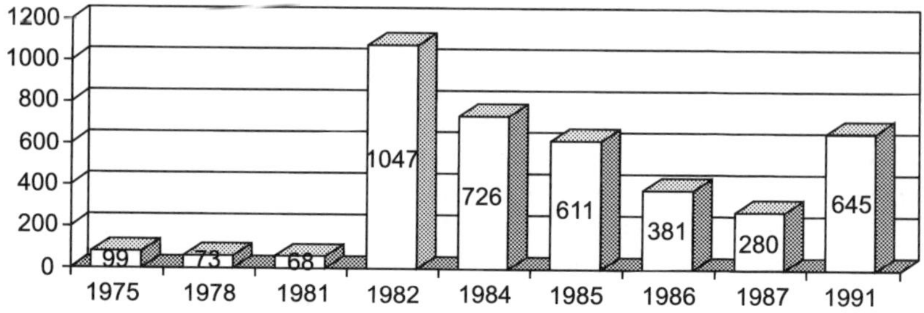

Figure 8.1 *Développement culturel*: *dépenses ordinaires* and *autorisations de programme*, 1975–91 (millions of francs indexed to 1991)

Source: *Ministère de la Culture*

Ministry's 'vertical', sectoral *directions*. However, the DDC's prerogatives *vis-à-vis* the other *directions* were never defined clearly and, as we shall see, the concept of 'transversality' was to prove extremely difficult to implement. More specifically, the DDC was also given the task of co-ordinating the Ministry's relations with the *collectivités locales*. Thus, it was responsible for negotiating a series of comprehensive funding agreements, either in the form of specific *conventions de développement culturel* or as part of the more general, inter-ministerial *contrats de Plan*, with individual towns, *départements* or regions. The Ministry thus committed itself to co-funding particular projects in the *collectivités* concerned, in return for a substantial financial commitment on the part of local policy-makers. In addition to these inter-sectoral tasks, the DDC pursued a more direct policy of supporting projects that appeared consonant with the Socialists' pre-election commitments, and which aimed, notably, at the recognition and encouragement of minority and 'community'-based cultural activities.

The DDC was the most important concrete expression of the Socialist Party's ideological commitments in the cultural policy arena[1]. Thus, in September 1981, the DDC's future director, Dominique Wallon, stressed its 'symbolic and political function as an illustration of an essential aspect of the new cultural policy'.[2] Similarly, Gournay (1988, p. 299) argues that

> the measures carried out by the new – and omnipresent – *direction du Développement culturel* and aimed at the least privileged categories of the population were doubtless the most typically 'Socialist' element of the policy unveiled after the Left came to power.

Jean-Pierre Colin, who served as adviser on *développement culturel* to Lang from 1981 to 1986 and again from 1988 to 1993, describes the DDC's output as 'the Ministry's most symbolic measures, [because] the conception of man and society that then prevailed is reflected in them' (Colin, 1986, p. 11). Urfalino (1993b, p. 845), emphasizing the legitimizing role of a policy of 'democratization' for the Ministry's cultural policy as a whole, suggests that the DDC was

> in a sense the administrative incarnation of the existence of a cultural policy, that is, of the resistance to an administrative segmentation that gives rise to an aggregate of sectoral *directions* entirely taken up by their relations with specific artistic or professional *clientèles*.

In view of the symbolic importance of the DDC, then, it is all the more significant that after the initial, expansionary phase of the early 1980s, the sector of *développement culturel* should have been marked by continuous institutional and budgetary decline. The DDC itself was abolished when François Léotard replaced Lang as Minister of Culture in 1986. Lang, on his return to the Ministry in 1988, did not, however, reinstate the DDC; instead, the sector of *développement culturel* continued to be the responsibility of a *sous-direction*, or 'sub-directorate', attached to the Ministry's central administration. In 1990, a more weighty *délégation au Développement et aux Formations* (DDF) was created – however, the DDF ranked below the status of a full *direction*, and its budget never reached the levels attained by the DDC in its first years. In fact, the decline of the DDC – and, by extension, of *développement culturel* as a policy priority – began several years before the DDC's actual abolition. The most obvious indication of this can be seen in the decline of the DDC's budget after 1982 (see Figure 8.1). To a certain extent, this downward trend is explained by 'one-off' programmes included in the 1982 budget, which were not carried over into subsequent budgets. However, none of the programmes funded by the DDC had the permanence of the institutions funded by other *directions* – it is thus difficult to distinguish between 'permanent' and 'ephemeral' lines in the budget of *développement culturel*. Moreover, even if we take account of items applicable only to the 1982 budget and subsequently discontinued, it is difficult to account for the continuous loss of credits between 1982 and 1986, which was much more pronounced than the modest reductions that affected most of the Ministry's other *directions* after 1983.

The DDC was not the only institution to suffer from a general decline of *développement culturel* after the first, expansionary years of

Lang's tenure as Minister. Until 1984, the *Fonds d'intervention culturelle* (FIC) continued to operate alongside the DDC, the policies of which it complemented with its explicitly interministerial projects. In the 1985 budget, however, the Ministry withdrew its funding for the FIC, which thus had to cease its activities. Officially, the FIC was abolished because it 'duplicated' the functions of the DDC. However, the decision provoked widespread criticism, both inside the Ministry and on the part of outside bodies like the Council of Europe[3].

In the remainder of this chapter, we will attempt to explain the evolution of Socialist policy towards *développement culturel* between 1981 and 1993. Specifically, we will consider, first, the creation of the DDC; second, the subsequent evolution of the administrative and budgetary status of the DDC within the Ministry, and its relations with other bureaucratic and political elements of the core executive; and third, the abolition of the *Fonds d'intervention culturelle*. We will begin by discussing the creation and subsequent institutional development of the DDC and its successors – not chronologically, but thematically, by considering the political resources that the DDC was initially able to draw on, and the factors that subsequently conditioned its decline.

8.2 THE CREATION OF THE *DIRECTION DU DÉVELOPPEMENT CULTUREL*

The establishment of the DDC can be explained by the conjunction, in 1981, of two discrete factors – a long-term partisan commitment and a short-term political calculation. The former, underlined by existing accounts of the Socialists' policy towards *développement culturel* (Saez, 1988; and Urfalino, 1993b), stemmed from the demand for a 'socialist' cultural policy expressed throughout the 1970s by influential sections of opinion within the Socialist Party. The latter, less obvious, factor relates to the unprecedented budgetary situation in which the incoming Minister found himself. Thus, Lang saw a structure like the DDC as a convenient way of accommodating a substantial proportion of the budgetary resources made available by the doubling, in 1981–2, of the overall budget of the Ministry of Culture. The ambiguity arising from the tension between these factors – the first relating to the balance of power within the Socialist Party, the second more contingent – contributes to explaining many of the difficulties subsequently encountered by the DDC. It will thus be expedient to consider, briefly, the interests and motivations of the different actors involved in the creation of the new *direction*.

The Creation of the DDC (I): the Partisan and Ideological Rationale

The creation of a *direction* responsible for the pursuit of a genuinely 'socialist' cultural policy – and for 'democratization' in general – had, throughout the 1970s, been one of the principal demands of the the cultural 'militants' within the Socialist Party. At an explicitly political level, these demands were articulated by the party's *Secrétariat national à l'action culturelle* (SNAC) – which, however, was dissolved in 1979, when its functions were transferred to Lang as *délégué national* for culture. More informally, the demand for a *direction* devoted exclusively to *développement culturel* was associated with a group of left-leaning civil servants and cultural administrators known as the 'clan des Grenoblois' because its members had all, at some point, been associated with cultural institutions based in Grenoble. After 1981, the 'Grenoblois' would assume many of the key positions in the DDC and other institutions concerned with *développement culturel*. Indeed, the 'figurehead' of the 'Grenoblois', Dominique Wallon, an *inspecteur des finances* and former president of the *maison de la culture* of Grenoble, who was close to the Rocardian, *autogestionnaire* wing of the Socialist Party, was appointed the DDC's first, and only, director. Wallon's appointment could be seen as the most explicit incarnation of the 'ideological', party political input into the Ministry's policy in this early period.

However, while the diffuse influence of the 'Grenoblois' within the Socialist Party may have created the conditions for an upgrading of *développement culturel* in the administrative hierarchy, we need to take account of more immediate and contingent factors in order to account for the precise institutional form taken by this general political objective. In the summer of 1981, it was by no means certain that the new administrative unit would have the full status of *direction*.[4] The Ministry had only one additional *directeur* post at its disposal; however, two separate 'emerging' sectors were competing for the new post of *directeur – développement culturel* and the visual arts. The former was defended most forcefully by Lang's first *directeur de cabinet*, Jacques Sallois, a former activist of the small left-wing *Parti socialiste unifié* who was sympathetic to Wallon's political outlook. The case for the visual arts was articulated by Lang's budgetary adviser, Claude Mollard.[5] The final decision may have owed as much to the distribution of power within Lang's *cabinet* as to wider political considerations. Indeed, according to a former *cabinet* member closely involved in the creation of the DDC, 'Sallois quite simply used the occasion of Mollard's going off on holiday to get the Minister to sign the document setting up the *direction*, the budgetary instructions, once the budget had been finished, and so when Mollard came back, he learned that he wouldn't be *directeur*.'[6]

The Minister played only a comparatively passive role in the decision: according to a senior official at the new *direction*,

> as far as the Minister was concerned, it wasn't his problem at all. He wasn't really interested in administrative and organizational problems, and as for the issue as such – the substance of the DDC's missions – let's say that he was totally ambivalent. He was both interested in certain aspects, and completely indifferent to others. So the Minister didn't really play a major role here.[7]

Instead, the prerogatives accorded the new *direction* owed much to the personal influence of the Minister's *directeur de cabinet*. In the words of one of Lang's closest advisers at the time, 'the DDC aspect, etc., and the choice of Wallon, and the concessions made to Wallon in the beginning, that was Sallois, for whom Wallon was a hero of the student struggle.... He was something of a symbol for someone like Sallois.'[8] The 'concessions' made to Wallon concerned, in particular, the attribution of the DDC's institutional competences, most controversially the *tutelle*, or administrative authority, over the 12 *maisons de la culture* and a network of around 50 smaller institutions that were co-funded by the *collectivités locales*. Until 1979, the *maisons de la culture* had been in the remit of the *direction du Théâtre et des Spectacles* (DTS); in 1979, they became the responsibility of the newly created *mission de développement culturel*. The decision to give the DDC control over these institutions – generally known as the *établissements d'action culturelle* (EAC) – was intended, officially, to provide the new *direction* with a conduit for imposing its aims 'on the ground' and throughout France. However, it was vigorously contested by the *direction du Théâtre* under its new director, Robert Abirached – who argued that it would be wrong to deprive the Ministry's sectoral theatre policy of a network of institutions which had, by then, increasingly come to resemble conventional theatres. Indeed, although the *établissements d'action culturelle* had previously been the responsibility of the *mission de développement culturel*, it seemed likely in 1981 that they would be transferred to the *direction du Théâtre*, given that Abirached had long been a close friend of Lang's, and that Lang viewed the notion of *développement culturel* with, at best, scepticism. As a senior member of Lang's *cabinet* points out, when having to decide in this dispute between the DDC and the *direction du Théâtre*, 'Lang himself wasn't entirely convinced, he hesitated a lot, and it's true that Lang, following his preferences, could very easily have decided otherwise.'[9] Lang's own retrospective assessment of the decision is even less ambiguous: 'The small mistake that I made, at the end of the day, is that, because of Wallon, I attached the *maisons de la culture* to the DDC.'[10]

The creation of the DDC, then, clearly owed much to partisan political considerations, and reflected the clout the proponents of *développement culturel* were still able to exert within the Socialist Party in 1981. However, as we have seen, purely contingent, personal factors, most notably the influence of Jacques Sallois within the *cabinet*, played an important part in overcoming internal bureaucratic obstacles to the creation of the DDC and the attribution of control over the *établissements d'action culturelle*.

The Creation of the DDC (II): the Budgetary Rationale

The political context described above may have been a necessary condition for the creation of the DDC. Whether it would have been sufficient to ensure the assent of Lang, who had always considered *développement culturel* secondary to the imperatives of artistic excellence, is a moot question. However, for Lang, the DDC was to serve another purpose which was quite distinct from the DDC's proposed 'democratization' policies, as initially outlined to him by Dominique Wallon in the summer of 1981, and for which the new Minister showed only modest interest. Lang thus saw the planned intersectoral *direction* that was to become the DDC as a financial 'reservoir' which would accommodate a large proportion of the funds made available by the near-doubling of his Ministry's overall budget in 1981–2. According to a senior member of Lang's *cabinet*,

> [Lang's] vision . . . consisted of saying, 'If this doubling of the budget that I've obtained is just distributed more or less in proportion to the respective budget of each institution . . . , it will disappear straightaway like water into the sands. Therefore, it would be wise to keep a kind of substantial reservoir, and the administration can take care of it, to launch new initiatives. . . . And these new ideas, we're not going

to have them all at once, so we'll need to keep this reservoir for a few years, which we can then empty progressively.'[11]

This vision did not necessarily coincide with the more 'programmatic' approach of the DDC's founders – in the words of a senior DDC official, 'Lang always thought there was money to spare there, and that whenever something was needed, a little outside the conventional rules, the DDC could be of use – which obviously then gave rise to conflicts with him.'[12]

In fact, there was fundamental disagreement between Lang and Wallon as to the ultimate *raison d'être* of the DDC. For Wallon, as we noted earlier, the new *direction* was to fulfil the 'ideological' objectives elaborated by the 'Grenoblois' in the 1970s. For Lang, by contrast, the DDC was primarily to concern itself not so much with systematic cultural 'democratization' as with 'innovation', a policy goal quite distinct from widening access in a socio-economic sense:

> For me – and I did hesitate, incidentally – the DDC was like a *direction* for innovation. I saw it as a *direction* that could be, shall we say, a *direction* for encouraging innovation, social innovation, or artistic innovation, in all the new areas in which we wanted to be active.[13]

As we shall see, the ambiguity which thus surrounded the aims of the DDC, and the ambivalence of the political support for its creation, would haunt the DDC until its abolition in 1986.

8.3. *DÉVELOPPEMENT CULTUREL*, 1981–93: THE VOLATILITY OF 'TRANS-SECTORAL' POLICY-MAKING

As suggested in the introduction to this chapter, the DDC was almost alone among the *directions* of the Ministry of Culture in not having a clearly circumscribed sectoral clientele. On the one hand, this potentially gave the DDC greater scope in defining its policy. On the other hand, it made the new *direction* highly dependent on the support of the political executive and on acceptance by the 'established' ministerial bureaucracy. Crucially, neither was unambiguously forthcoming.

The 'Horizontal' Vocation of the DDC: the Limits of Non-sectoral Policy-making

The DDC was established with the explicit aim of acting as a catalyst for a more innovative policy style that would promote the objective of cultural 'democratization' across all of the Ministry's areas of intervention.[14] In fact, the gulf between this ambition on the one hand, and the lack of corresponding statutory powers or political resources on the other, accounted in large part for the DDC's growing difficulties between 1982 and 1986. In certain policy fields, the DDC was given clear responsibilities – notably with regard to the transfer of funding in the context of the *conventions* with sub-national government. However, the DDC had no powers to impose any form of cross-sectoral co-ordination within the Ministry; nor did it have the power to impose a 'democratization' strategy on the sectoral *directions*. Furthermore, in spite of Wallon's efforts to this end, it failed to obtain outright control over the Ministry's field units, the *directions régionales des Affaires culturelles*.

Perhaps most important, problems arose whenever the DDC intervened directly in activities which could be construed as falling into the remit of one of the traditional *directions*. The DDC's relations with Robert Abirached's *direction du Théâtre* (DTS) were particularly strained. There were several potential areas of conflict between the two *directions*. Abirached resented the DDC's attempts to establish itself as the natural partner for theatrical projects, both amateur and professional, with an 'ethnic' or minority dimension, which he saw as confining his *direction* to a narrow institutional 'ghetto': 'There was almost a tendency to say, "Right, you're doing theatre, you're Armenian, or you're Basque – go and see the people at the DDC, they've got funding for all that." But this became completely unacceptable, in a country in which there is a "global" theatre policy.'[15] On numerous occasions, then, there were demarcation disputes between the DDC and the DTS over the funding of specific projects. In most cases, these were resolved in meetings between Wallon and Abirached. Where disagreement persisted, the issue would be referred to an arbitration meeting at the level of Lang's *cabinet*, in which the *directeur de cabinet* – Jacques Sallois, for most of the period concerned – adjudged the competing claims of the two *directions*. In exceptional cases, the issue would be submitted for a final *arbitrage* to the Minister himself – where, in the words of one senior DDC official,

> it was apparent that Abirached was much more successful at having the Minister's ear than Dominique Wallon, and that Abirached generally went on the offensive in this way to make up lost ground. On the other hand, Dominique Wallon's relationship with the Minister became very fragile at one point.[16]

Moreover, relations between the DDC and the DTS were rendered increasingly tense by the persistence of disagreement over the future of the *établissements d'action culturelle*. Far from being in a position to use the EAC as an instrument of a coherent 'democratization' policy, the DDC found itself having to manage the gradual transition of the *maisons de la culture* from their original function of 'animation' to a more narrowly artistic vocation, which was vociferously demanded by the directors of the EAC and their professional organization, the SYNDEAC[17]. The DDC was thus destabilized by a task which Dominique Wallon considered central to his 'mission' – not least in view of his personal background as President of the *maison de la culture* of Grenoble – but for which even some civil servants at the DDC considered the new *direction* ill-equipped. According to a senior DDC official, 'it was a managerial job in an area which had nothing to do with the objectives of an administration geared towards "missionary" tasks, the DDC'.[18] The DTS, moreover, had not surrendered its hopes of eventually regaining control of the *maisons de la culture*. In the words of a DDC official, 'Abirached worked at undermining [the work of the DDC], and had a great time playing tricks [on the DDC]. When [Dominique Wallon] made a proposal to Lang for a decision concerning a *maison de la culture*, for appointing a director, Abirached came along with his own candidate from the theatre world, and so on.'[19] The DTS was, in fact, to regain control over the EAC after the DDC's abolition in 1986; in 1990, it was to accomplish the restructuring of the sector with the introduction of the label *scènes nationales* for the former EAC to underline their essentially artistic orientation.

Jack Lang and the DDC: a Difficult Cohabitation

The pattern that characterized the relationship between the DDC and the DTS was equally applicable, though less acutely, to relations with the other traditional *directions*. Perhaps just as important in explaining the progressive weakening of the DDC, however, was its gradual loss of support at the ministerial and *cabinet* level. The departure of Jacques Sallois as *directeur de cabinet* in 1984 deprived Dominique Wallon of an important personal ally. Indeed, according to a senior DDC official,

> after the departure of Sallois..., things let up, and the *cabinet* became more and more hostile to Dominique Wallon and the DDC. And after that, for two years, there was an extremely tough

> and disagreeable power struggle, and one had the sense that the *cabinet* saw the maintenance of the DDC in the Ministry as an injustice.[20]

The sectoral *conseillers* in the *cabinet* were, predictably, defending the interests of their respective sectoral *directions* – which inevitably jarred both with the DDC's increasingly unrealistic aspiration to a 'supra-sectoral' status and with its use of funds which the other *directions* would gladly have assumed themselves.

The tensions between the DDC and the sectoral bureaucracy, which thus found expression both at the level of the *directeurs* and within the *cabinet*, also influenced the attitude of the Minister himself. It had always been Lang's natural inclination to support the claims of 'création' against the less tangible benefits promised by *développement culturel*. Yet while Lang was initially forced to accommodate the proponents of *développement culturel*, the political context had, by 1984, changed in important respects. Lang had consolidated his position within the government, both as a result of his close relationship with the President and in virtue of his unrivalled personal popularity. Concomitantly, the role of the Socialist Party, and of Socialist deputies in particular, in determining the priorities of the Ministry of Culture had faded into insignificance – a situation which mirrored the growing impotence of parliament in other areas of policy. In part, this was explained by the fact that, as the 1986 elections approached, the attention of deputies inevitably shifted to immediate electoral concerns – in the words of a senior member of Lang's *cabinet*, 'in 1986, they didn't know the DDC any more. All they saw was their constituency – they were no longer available for this kind of debate.'[21] At the same time, as the same *cabinet* member points out, Lang had always derived his ultimate legitimacy as Minister from his links with the Elysée: 'Lang was a man of the court, and so he wasn't going to engage in long meetings with ordinary members of parliament. It wasn't his level.'[22]

The DDC's position in the Ministry was further weakened by a deterioration in the personal relationship between Lang and Dominique Wallon. In the words of a *cabinet* member closely acquainted with both Lang and Wallon, 'Dominique Wallon wasn't very close to Lang – they never really got on .... Wallon represented the Rocardian wing of the party, he was anything but a friend of Lang's, and their personalities were very different.'[23] As suggested earlier, Wallon's political outlook diverged markedly from Lang's: Wallon stressed the importance of long-term educative work, of cultural decentralization and of strengthening a network of grassroots *associations*, while Lang emphasized the absolute priority of *création*, favoured visible and spectacular policy initiatives and rejected functional decentralization in the sense indicated by the Defferre reforms. However, the tension between Lang and his *directeur du Développement culturel* was to crystallize around one specific issue – the Ministry's policy towards the *établissements d'action culturelle* (EAC). As we have seen, Lang decided only with great reticence to attribute the *tutelle* of the EAC to the DDC. Subsequently, according to a senior DDC official, 'Lang reminded people from time to time that ultimately, it wouldn't have been so bad [to attach the EAC to the *direction du Théâtre*], that he wanted the *créateurs* to be there, that the *créateurs* meant the theatre, etc.'[24] The conflict is presented by Lang himself as a fundamental clash between the conservatism of the DDC and his own desire to reform the EAC:

> The DDC was very innovative in a great number of areas, and very conservative in the case of the *maisons de la culture*. This kind of costly, inert bureaucracy – it took me many years before I could change that, when in fact I wanted to do it very quickly. . . . Because these *maisons de la culture* were ossified and bureaucratized to an unimaginable degree. So I waged permanent war against the DDC, it's true, and I told Wallon all the time, 'But you're not doing anything, you're promising a reform all the time, and nothing is happening.'[25]

At the level of specific policy decisions, the tension between Lang and Wallon, particularly in the second half of the 1981–6 legislature, was centred on appointments to individual *maisons de la culture*, where Lang imposed 'artistic' directors against Wallon's opinion. In one instance, the difference of approach between Lang and Wallon almost led to the latter's resignation. Thus, in November 1984, Wallon threatened, in writing, to resign over a disagreement with Lang which ostensibly concerned the *maison de la culture* of Le Havre, but which, in the words of a senior member of Lang's *cabinet*, was 'more about the general relationship between Lang and Wallon than about any objectives over which they might have openly disagreed'[26]. For several weeks, Lang and Wallon did not meet, even in the formal context of the habitual monthly meetings between the Minister and his *directeurs*. The conflict was only defused by the personal mediation of Lang's then *directeur de cabinet*, Thierry Le Roy. The *maisons de la culture* may have been the manifest cause of the tension between Lang and Wallon;

on the other hand, the conflict increasingly spilled over into other areas of policy – and more generally, it seems plausible that this personal factor affected budgetary decisions relating to the DDC.

The Budget of the DDC: Ministerial Reservoir' or Policy Instrument?

As we have seen, the DDC was initially accorded a volume of credits which, in budgetary terms, placed it on a par with some of the established sectoral *directions*. From the 1983 budget onwards, the allocations set aside for *développement culturel* diminished rapidly – by around 12 per cent in real terms from 1982 to 1983, by around 11 per cent from 1983 to 1984 and again from 1984 to 1985, before plummeting dramatically as a result of the change of government in 1986. Significantly, while almost all *directions* had to accept limited budgetary regression after 1983, the decline in absolute terms was accompanied, in the case of *développement culturel*, by a fall in the sector's share of the overall budget of the Ministry of Culture – from 12.1 per cent in 1982 to 7.3 per cent in 1985. It is of course impossible to provide a comprehensive explanation for this decline, which would necessitate disaggregating countless individual decisions by Lang and his *directeurs de cabinet* on specific budgetary lines in successive ministerial budgets. However, a number of factors which may account for this phenomenon have already been cited. The general weakening of political support for the DDC and the troubled relations between Lang and Wallon were discussed at some length in the preceding section. To these, we should further add the conception of the DDC as a 'reservoir', useful in absorbing a substantial proportion of the budgetary increase of 1982, and which could subsequently be 'emptied'. Following this conception, the DDC had been given the task of administering a number of 'one-off' programmes, specially conceived by Lang's *cabinet* as a response to the challenge of the 1982 budget, and which were gradually phased out in subsequent years. The most important of these, with a budget of around 200 million francs, involved the funding, by the Ministry, of a 'job creation' programme in the cultural sector – described by a senior DDC official as 'a digression . . . which the DDC managed as best it could, and then we progressively pulled out of it'[27].

The status of the DDC's budget – or rather, to be more precise, of the lines in the Ministry's budget controlled by the DDC – was thus never defined without ambiguity and in identical terms by Lang and by the DDC itself. For the former, it was to serve at least in part as a 'reservoir' that could be depleted as circumstances dictated; for the latter, it corresponded to specific programmes which were intended to have a certain stability over time. This fundamental contradiction could be seen as reflecting the underlying problem of the DDC – an insufficient fit between the ambitions and policy agenda of its proponents on the one hand, and the definition of a political programme, and political support, on the other. A similar pattern would characterize the Ministry's policy during the 1988–93 legislature.

1988–93: the *Manquée* Rehabilitation of *Développement Culturel*

The DDC was abolished by François Léotard, Lang's successor at the Ministry of Culture, a few months into the first *cohabitation*, and its responsibilities were transferred to the Ministry's *direction de l'Administration générale*, which was renamed *direction de l'Administration générale et de l'Environnement culturel* (DAGEC). Concomitantly, the *établissements d'action culturelle* were attached to the *direction du Théâtre*, and the credits of *développement culturel* were reduced drastically. In fact, according to one senior DDC official, the DDC was already 'dying... and profoundly weakened from within' when the Right came to power in 1986.[28] Even before the 1986 elections, Lang himself had been contemplating a reform of the Ministry's internal structures – and this might well have entailed the abolition or downgrading of the DDC. According to Lang, such changes would have helped to 'protect' the budgetary increases obtained for the Ministry in 1982 against the incoming conservative government: in Lang's own words, 'what I regret is that in 1985, in 1986, I didn't redistribute the funds of the DDC among the different *directions*, because they would have been preserved more effectively by the artists'.[29]

However, when Lang returned to the Ministry in 1988, the DDC was not re-established; nor were the EAC attached to the *sous-direction* that had succeeded the DDC. Nonetheless, in February 1990, a *délégation au Développement et aux Formations* (DDF) was created, headed by Hélène Mathieu, who had until then been *conseiller technique* for *développement culturel* in Lang's *cabinet*. Although the policies of the DDF need not concern us here, it will be worth briefly considering the circumstances of its creation, and its relations with the Minister, *cabinet* and administration, as these bear out once again the structural constraints on a non-sectoral 'democratization' policy already discussed with reference to the DDC. Moreover, policy towards *développement culturel* in the 1988–93 period points even more strongly than the history of the DDC to the crucial importance of personal factors – that is, the relationship between the Minister and specific civil servants – as a determinant of policy in this sector.

The creation of the DDF: personal and organizational factors

When Lang returned to the Ministry of Culture in 1988, there was little enthusiasm on the part of either the Minister or his *cabinet* for re-establishing the DDC. Dominique Wallon and the 'Grenoblois', conversely, were not inclined to repeat the experience of the DDC in the absence of adequate political support. Moreover, the political context had changed; what little party political or ideological input into the Socialists' cultural policy priorities there might have been in 1981 had, by now, faded. However, while the status of *développement culturel* was not a priority for either Lang or his new *directeur de cabinet*, both were determined to change the status of the 'super'-*direction* created by Léotard, the DAGEC – partly because of the right-wing sympathies of its then director, and partly because the sectoral *directions* were concerned about its excessive weight in the Ministry's hierarchy. Initially, the *cabinet* contemplated simply placing, in the words of one of Lang's advisers, 'one of us' at the head of the DAGEC, thus avoiding the complications of restructuring the Ministry yet again:

> We would have saved ourselves the reform of the administrative structure, because in addition, in order to change the structure, you need to go to Matignon, it's complicated, you need the agreement of the Elysée, it's a very, very heavy procedure. So, if we could have got round it, we wouldn't have undertaken this administrative reform.[30]

However, the *cabinet* could not agree on a suitable candidate; at the same time, Lang's *conseiller technique* for *développement culturel*, Hélène Mathieu, was 'lobbying' the Minister to create a separate *direction* to administer the sector. The conjunction of these two factors appears to have persuaded Lang: in the words of a senior *cabinet* member, 'we said to ourselves, "Well, actually, we've got the right person to take care of this, who really wants to make an effort in this area", and so we created the structure. It really is linked to considerations relating to particular individuals.'[31] At the same time, the creation of a separate *délégation* for *développement culturel* was opposed by Lang's *directeur de cabinet*, Francis Beck, who delayed Mathieu's appointment for several months – according to a *cabinet* source, 'the very day that [Hélène Mathieu] was appointed in the Council of Ministers, Francis Beck ... called Matignon and the Elysée, saying, "Above all, don't appoint her!" '[32] Furthermore, not least in view of the impressions left by the DDC, the sectoral *directeurs* viewed the creation of another 'horizontal' *direction* with scepticism. The Minister himself was ambivalent about the new institution, which promised to deprive the sectoral *directions* of some of their budgetary allocations: according to a *cabinet* member closely involved in the creation of the DDF, '[Lang] didn't want the DDF.... And even today, he is convinced that it was foolish, that one should have put these funds where the artists are, because the artists would have gone out to demonstrate in the streets as soon as one had tried to cut the funding.'[33] Indeed, in the course of elaborating the 1993 budget, Lang contemplated redistributing the credits of the DDF to the sectoral *directions*. As in 1986, when a change of government similarly appeared inevitable, the Minister thus hoped to save a component of the budget which appeared a likely target for cuts under the incoming government. In Lang's own words,

> I realized very well that if the Right came to power, they might cut funding – and what were they going to cut first? The least visible kinds of funding. The funds that relate to the provinces. [Hélène Mathieu] said, 'No, no, it will be protected by the local politicians.' But the local politicians don't do anything – they'll just take it lying down. It's better to put the funds under the protection of the artists – who will perhaps fight. Once again, I gave in to the kindness of the person in charge – at first it was Wallon, then it was Hélène Mathieu ...[34]

In short, the institutional evolution of *développement culturel* between 1990 and 1993 again serves to underline the extent to which the sector was subject to contingent personal factors. While relations between the Minister and the head of the DDF contrasted with the acrimony between Lang and Wallon which had contributed to the decline of the DDC, the status of the DDF within the Ministry was characterized by institutional and budgetary fragility not unlike that of its precursor in the early 1980s. The Minister continued to be sceptical with regard to the notion of *développement culturel*, and remained detached from the elaboration of the DDF's concrete policy output. Moreover, both the creation of the DDF in 1990 and Lang's decision not to transfer its credits to the sectoral *directions* in the 1993 budget followed directly from the personal influence of Hélène Mathieu and the close relationship between Mathieu and Lang – which, in the absence of overriding political considerations or a cohesive sectoral policy network, could be seen as the central determinant of policy towards *développement culturel* after 1990.

8.4. THE LANG MINISTRY AND THE *FONDS D'INTERVENTION CULTURELLE*

The DDC was, as we have seen, the main institutional conduit for *développement culturel* from 1982 to 1986. However, our analysis of the tensions and weaknesses inherent in the Ministry's policy in this sector is further strengthened by an account of policy towards a smaller institution which played a crucial role in propagating *développement culturel* until the early 1980s – the *Fonds d'intervention culturelle* (FIC). The FIC was created in 1971 on a recommendation of the VIth Plan;[35] in a decree defining its legal status, the FIC's general objective was subsequently defined as 'enabling the launch of innovative or experimental operations which are jointly elaborated, implemented and funded by several government departments or by central government and the *collectivités locales*' (quoted in Gournay, 1988, p. 359). The FIC was formally placed under the authority of an inter-ministerial committee and financed by interministerial funds; in practice, the general secretariat of the FIC was closely integrated with the administrative structures of the Ministry of Culture. However, the methods of the FIC were in stark contrast with the funding mechanisms prevailing in other sectors: no project could be supported for more than two years; the FIC's support could only be used to cover running costs or materials; and the FIC's grant was not allowed to exceed 50 per cent of a project's total cost. In the decade following its creation, the FIC supported around 1,200 projects; yet the sum of the grants made over this period did not exceed 140 million francs. As a report on French cultural policy by the Council of Europe noted in 1988, 'even if one takes the multiplicatory coefficient into account, this sum is derisory when compared with institutional spending on high culture' (Wangermée, 1988, p. 213).

Its modest resources notwithstanding, the FIC was 'abolished' when the Ministry of Culture failed to renew its contribution to the FIC's budget, a sum of around 15 million francs, in the course of elaborating the 1985 budget. The official rationale for this decision pointed to the creation of the *direction du Développement culturel* with, ostensibly, very similar aims to those of the FIC. However, the decision to scrap the FIC was far from uncontroversial. The Council of Europe's *rapporteur* described it as 'regrettable' (Wangermée, 1988, p. 213). Jean-Pierre Colin, responsible as *chargé de mission* for *développement culturel* in Lang's *cabinet* throughout the first Lang Ministry, was fiercely critical of the decision, observing that 'the disappearance of an instrument which was so manifestly adapted to the Ministry's new policy, and which had only recently been improved in its functioning, was a shadowy affair' (Colin, 1986, p. 56). Nor was the decision supported unanimously by the civil servants in the DDC itself – according to one senior DDC official, the abolition of the FIC was 'incomprehensible' and 'there was always a need for a real small *force de frappe* to deal with cases that were out of the ordinary, that had an inter-ministerial aspect'.[36] Whether or not these criticisms were justified, the decision to abolish the FIC yields an interesting case study of decision-making and the power relations between different actors in the sector of *développement culturel* – particularly in view of the symbolic significance of an institution which, both in its aims and in its methods, diverged from the structures and organizational patterns prevalent in the 'traditional' sectors supported by the Ministry. It is not, of course, the aim of this section to 'evaluate' the decision to dismantle the FIC, which may or may not have been a necessary concomitant of the creation of the DDC. However, the starting point for our discussion is the absence of a consensus among different actors within the administration about the role of the FIC, and the consequent need to explain why the argument for the FIC's abolition prevailed over those advocating its preservation.

The Abolition of the FIC: a Case of 'Technocratic' Decision-making?

The creation of the DDC in 1982 undoubtedly cast a question mark over the future of the FIC, the general aims of which could now be said to be duplicated by those of the DDC. Nonetheless, it was felt that the FIC still had a purpose, not least by Dominique Wallon himself. In the words of a senior DDC official, '[Wallon], at the time, was arguing in favour of keeping the FIC, because, he said, one could not be sure what exactly the DDC would be doing, and because it was desirable to maintain a structure that would play the role of a "free electron"'.[37] However, Wallon also felt that the tasks of the FIC needed to be redefined to take account of the parallel existence of the DDC. This view was shared by the *cabinet* members most directly concerned, Jacques Renard, who followed the sector as *conseiller technique*, and the *directeur de cabinet*, Jacques Sallois. In the winter of 1983–4, Wallon and the FIC's new secretary-general, Anne Jullien, jointly developed a set of proposals for a 'new' FIC, which were submitted to, and discussed with, Renard and Sallois – however, in the words of one senior DDC official, 'it is clear ... that a merger of the FIC and the DDC was one aspect of what the *cabinet* and Wallon had in mind'.[38]

Crucially, this 'réflexion' about the relationship between the FIC and the DDC occurred in a period of increasing budgetary pressure on the Ministry of Culture as a whole, to which, as we have seen, the sector of *développement culturel* was particularly vulnerable. The decision to abolish the FIC altogether was then taken – without public debate or consultation and, according to a senior DDC official, 'surreptitiously'[39] – by Renard and Sallois in the summer of 1984. The justification put forward for this decision was one of inevitable 'rationalization': in the words of a senior member of the *cabinet* closely involved in the decision, 'the FIC was abolished because it had been successful' – in other words, because it had inspired the newly established DDC and the Ministry's regional *directions* to develop similar activities. Moreover, according to the same source,

> for several months, we tried to say . . . : 'Right, as the FIC has played its role of encouraging, and as the DDC now exists, what can be the FIC's new identity?' . . . We told the people in charge of the FIC: Try and spell out a new FIC for us! Give us some practical and credible missions for the FIC! No one was capable of identifying what the FIC's new missions might be, and they weren't because it wasn't possible.[40]

At the same time, the FIC's disappearance was welcomed, for other reasons, by a number of influential groups within the administration. The Ministry of Finance, supported by the Prime Minister's office, had long been pressing for an end to a mechanism which, by definition, relied on contributions from a large number of ministries and agencies. As a senior member of Lang's *cabinet* recalls, 'the Ministry of Finance asked that the FIC be abolished – because it was an inflationist system, such that when you put in one [franc], you obtained five. You got the different ministries to cough up money, and so it was an infernal instrument for the Ministry of Finance.'[41] Moreover, the attitude of the sectoral directorates of the Ministry of Culture resembled their scepticism *vis-à-vis* the DDC, as a former FIC official points out: 'The FIC had always annoyed the other directorates inside the Ministry of Culture, and so a lot of *directeurs* weren't all that displeased when it came to a bad end.'[42]

To conclude this brief account of policy towards the FIC, then, we may note that, while the FIC's role was undoubtedly affected by the creation of a *direction du Développement culturel*, the decision to abolish the institution has to be seen against the background of a number of factors which weakened its position within the Ministry.

Writing in 1982, before the disbandment of the FIC, Djian (1982, p. 371) had already pointed to the 'institutional and political instability' which resulted from the fund's inter-ministerial and inter-sectoral character. As we have seen, the 'instability' observed by Djian was compounded, notably, by the opposition of the Ministry of Finance and of the established *directions*, and, conversely, by the lack of politically significant sources of support either within or outside the Ministry of Culture. These sources of weakness closely mirror the difficulties encountered by the DDC, which were considered earlier.

We are now in a position to discuss the wider implications of the patterns of policy-making encountered in the field of *développement culturel*; in particular, we can consider the contribution the policy network approach might make to understanding the status of an 'intersectoral' approach in a policy arena dominated by clearly delineated sectoral policies.

CONCLUSION

This chapter has focused on three factors affecting policy towards *développement culturel* in the 1980s. First, the relevance of party political and ideological influences, initially decisive, declined progressively after 1981. Second, 'personal' factors, notably the relationship between Lang and the directors of the *direction du Développement culturel*, and later the *délégation au Développement et aux Formations*, had a demonstrable impact on the strength of ministerial support for the Ministry's efforts in the area of *développement culturel*, not least in budgetary terms. Third, the DDC, the FIC and, to a lesser extent, the DDF all encountered a significant degree of hostility on the part of the Ministry's traditional *directions*. However, while these factors could be seen as necessary conditions for the dramatic fluctuations in the institutional status of *développement culturel* observed in the 1980s, they do not in themselves offer a sufficient explanation. In order to provide a comprehensive characterization of the structural constraints on policy-making in this sector, and to emphasize the specificity of *développement culturel* when compared with other sectors, we have to consider a fourth, structural variable – the ability, or inability, of an organization like the DDC to draw on the support of actors and organizations outside the core executive. To elaborate this final, and crucial, point, it will be useful, as in previous chapters, to draw on the policy network approach.

Thus, with regard to previous case studies, it was possible to establish a clear correlation between the nature and stability of policy outputs on the one hand, and the existence of policy networks and strength of professional interests in these networks, on the other. The contrast with *développement culturel* could not be more marked. Applying, once again, the typology developed by Rhodes and Marsh (1992, p. 187), the loose policy-making system concerned with *développement culturel* conformed closely, on most dimensions, to the 'issue network' ideal type identified by Rhodes and Marsh. Thus, the network comprised a large range of diverse interests; it was very loosely integrated and there was frequent change in the intensity and frequency of contacts between different members of the network. With regard to the degree of consensus within the network, identified as a crucial dimension and determinant of networks by Rhodes and Marsh, we found that there was only limited agreement among different actors about policy aims; policy towards the *établissements d'action culturelle*, in particular, was periodically subject to intense conflict between key actors in the policy-making system. Moreover, with regard to the distribution of power resources between different organizations, the system differed in crucial respects from the networks described in previous chapters. The degree of interdependence between members of the network – defined by Rhodes (1981, 1985) in his 'power dependence' framework as based on a necessary exchange of resources between organizations – was limited. Indeed, the relationship between public and private actors in the area of *développement culturel* was essentially consultative. Patterns of policy-making thus conformed much more closely to a 'hierarchical' ideal type than in most other sectors considered; conversely, the ambivalence and instability of decision-making by the political segments of the core executive had a more profound and immediate impact on *développement culturel* than on any of the 'traditional' sectors.

Développement culturel and the Wider Cultural Policy Network

So far, we have treated the policy-making system concerned with *développement culturel* as a self-contained network, albeit a very loose one. Alternatively, we might conceive of *développement culturel* as part of a wider issue network relating to the policy output of the Ministry of Culture in general. The salient feature of a comprehensive national cultural policy network would lie in the distribution of power resources between different groups of actors. In the Rhodes and Marsh typology, the ideal-type issue network is characterized by unequal power distributions within the network, reflecting an unequal spread of resources and access; moreover, the issue network is characterized, in contrast with the 'policy community' ideal type, by the prevalence of zero-sum outcomes. There is a clear fit, then, between the kind of constellation described by Rhodes and Marsh and the account sketched in this chapter of the status of *développement culturel* within the overall 'cultural policy' pursued in France between 1981 and 1993. However, the implications of this analysis concern not only the specific sector of *développement culturel*. Rather, using *développement culturel* as a 'controlling case' diverging from the sectoral paradigm observed in previous chapters, we can now proceed to establish more general inferences about the variables that affect policy outcomes in the area of cultural policy.

Conclusion

GOVERNING THE ARTS IN FRANCE: PROFESSIONALS, NETWORKS AND THE STATE

In Chapter 1 of this book, we noted a paradox. In keeping with the monarchic and Jacobin traditions which underlie the organization of institutional relations between the state and society in the Fifth French Republic, the French arts funding system is highly centralized and subject to the direct authority of the political executive; indeed, cultural policy is frequently seen as having moved into a presidential *domaine réservé*. Moreover, France is unique among liberal democracies in the extent to which cultural policy is invested with an ideological 'mission': in the words of Saez (1993, p. 69),

> French political culture, which we could also call the 'Republican model', is based on a system of integrationist values; this system has established participation in culture, under the control of the state, which thus acts as the guarantor of democratization, as one of the most important ways of acceding to citizenship.

The period of 1981–93, which saw the Left exercising undiluted power for the first time since the Popular Front, might have been expected to exemplify these precepts. However, at the prosaic level of concrete policy outputs, the policy pursued by the Ministry of Culture was somewhat at odds with the 'voluntaristic' republican ideal. The Socialists' ascent to power did not herald a radical shift in the Ministry's priorities, which have traditionally emphasized the promotion of artistic 'excellence'. The underlying pattern of policy was one of continuity, though with greatly enhanced budgetary means. A more detailed analysis of policy and policy-making with regard to six selected sectors or projects has confirmed this pattern – our last case study, devoted to *développement culturel*, serving to underline the limits of the overtly political, 'ideological' strand in Socialist cultural policy.

How can we account for this paradox? From the evidence presented in our case studies, it is clear that the key to understanding the peculiar tension between the proclaimed 'voluntarism' of French cultural policy and the fundamental continuity of policy outputs has to be sought in the relationship between public officials and their professional 'clients'. In order to identify the underlying components and dynamics of this relationship, we will proceed in three stages: we begin by considering the extent to which public and private actors, respectively, were able to shape the elaboration and setting of policy in different sectors; we then move on to consider the bases of professional influence in individual sectors, and the consequences of professional power for the overall shape and direction of cultural policy; finally, we consider the wider implications of our findings, both with regard to general 'theories' of cultural policy-making in liberal democracies, and for the study of state – society relations in France.

CULTURAL POLICY-MAKING, 1981–93: SECTORAL NETWORKS AND THE LIMITS OF STATISM

In each of our sectoral case studies, we sought to answer a simple question: who 'governs' cultural policy-making in France? The views which we have described, throughout this book, as 'institutionalist' and 'statist' would suggest a very simple answer to our question. The formulation of policy, on the institutionalist view, is the preserve of a clearly delineated set of actors occupying institutional positions within the state, or within a more narrowly defined cluster of institutions, such as the 'core executive'. An answer to the question 'who governs?', according to this approach, would consist simply in identifying the subsets of institutional actors who contribute to shaping particular policies. Institutionalist interpretations are tempting in their simplicity, and we noted that existing accounts of French cultural policy-making have tended to favour variants of the institutionalist view – usually in the form of 'presidentialism' or of 'ministerial government', or of a combination of the two. Yet clearly, none of the sectors considered in this book conforms closely to either the presidential or the ministerial type of decision-making, or to any of the other core executive 'models' identified by Elgie (1993), discussed at the beginning of this book. At the very least, we would argue, on the basis of our sectoral case studies, that the role of intra-governmental actors, whether elected politicians or bureaucrats, cannot be conceived of in isolation from extra-governmental influences on the policy process. In almost all of the sectors considered, significant policy decisions were, almost without exception, the result of close collaboration between public actors and their professional 'clients'; in fact, in many instances, particular policy initiatives clearly originated not with elected politicians, their advisers or even civil servants, but with individual professionals or groups of professionals. At the same time, the 'clients' of the Ministry of Culture acted in many sectors as effective breaks on the ability of public actors to impose policy change where such change would have threatened the interests of the professionals concerned.

Moreover, the pattern of interaction between public actors and professional interests which we observed in our case studies amounted to more than merely the occasional participation of professionals in the policy process. Professional collaboration with public policy-makers was not limited to 'consultation' on particular technical policy issues; nor was professional influence over the policy-making process exhausted by the ability of professionals to block or modify proposals emanating from within the core executive. In most of the sectors considered, and in all of the 'core' sectors of public funding, policy formulation was the preserve not of any particular actor within the core executive, or even of a fluctuating group of actors within the core executive, but of a set of actors straddling the divide between the public and private sectors. In short, we situated the locus of policy-making not within the core executive, but in policy networks centred around sectoral funding mechanisms, and characterized, to varying degrees, by stability and continuity in their composition and in the intensity of contacts between different members.

Yet to observe that policy was generally formulated in collaboration between public officials and cultural professionals is not to deny that public actors played a decisive role in many of the sectors under consideration, or to suggest that different 'models' of core executive operations are irrelevant to our analysis. Thus, while the policy network approach offers an image of policy-making more accurately accorded with our empirical evidence than an isolated focus on either the core executive or professional interest groups, we also noted that there was a substantial degree of variation among the sectors considered with regard to the influence exercised in the policy-making process by 'institutional' actors such as the Minister of Culture or civil servants. Some networks were clearly characterized by a balance of power between officials and cultural professionals; in other sectors, there was a clear bias in favour of public actors. Before we consider any of the other dimensions of the cultural policy networks which we were able to identify, we must first consider, therefore, the differences in the degree of 'state' influence across sectors, and suggest explanations for the variance. Concomitantly, this will enable us to specify more clearly the relevance of institutionalist explanations of the policy process, and to clarify the relationship between narrowly institutional accounts and the policy network approach.

Policy Networks and the Core Executive: From Dichotomy to Continuum

As we suggested earlier, the policy-making process was characterized in all of the sectors considered by close collaboration between officials and professionals; yet the respective role of public and private actors differed markedly from sector to sector. The differences evidently call for explanation; and as a first step towards an explanation, we need to produce an ordering of networks according to the dimension identified here. None of the theoretical classifications of network types which we considered at the beginning of this book explicitly introduce a measure of the respective influence of public and private actors within networks. Rhodes and Marsh (1992, p. 199) suggest that the more closely integrated kinds of network – that is, 'policy communities' – are likely to be composed of a combination of public and private sector elites: 'Producer groups and professional groups are the groups which, together with the government, dominate the policy networks.' The Rhodes and Marsh framework does not, in other words, attach great significance to the distribution of power resources between professional interests and government: public actors are assumed to 'side' with a prevailing professional or producer interest. Yet considering the specific sample of empirical cases studied in this book, we clearly need to account for a more diverse range of possibilities. We thus propose to introduce a dimension of comparison focused specifically on the degree to which the state – or, more neutrally, public actors – influence what Rhodes (1981, p. 122) terms the 'appreciative system' prevailing in a given policy network. Specifically, we would suggest that the cases we examined can be ordered on a continuum of networks ranging from a 'statist network' at one end to a 'professionalized network' or 'professional network' ideal type at the other. A statist network is defined simply as a network in which the policy agenda or 'appreciative system' is determined, and effective policy choices are set, by public officials – a category of actor which may, for these purposes, include any of the members of the core executive. It should be noted that a 'statist network' ideal type should be seen as distinct from plain 'statism' – which, as Van Waarden (1992, p. 42) points out, implies 'strong state intervention but without the involvement of societal actors', such that 'the state consciously tries to exclude organized interests from policy making and policy networks'. By contrast, a statist network in the sense proposed here implies that societal interests are included in policy-making – yet always subject to the controlling influence of public actors. With regard to the concept of professional network, we may follow the established usage of the term: Rhodes (1985, p. 19) thus defines professional networks as 'characterized by the pre-eminence of one class of participant in policy-making – the profession' and as 'express[ing] the interests of a particular profession and manifest[ing] a substantial degree of vertical independence whilst insulating themselves from other networks'. At one end of the continuum, the interests of public actors prevail, while at the other, professional interests are dominant.

Applying this simple typology to our sectoral case studies serves to underline the very real variance between cultural policy sectors on the dimension of 'state' influence. Perhaps more important, it produces results which may appear surprising, given the 'manifest image' of cultural policy-making in France, and most notably, the cliché of presidential 'activism' in the cultural policy area. The one 'presidential' *grand projet* included among our case studies, the *Cité de la Musique*, is thus located not at the 'statist network' end of the continuum, but towards the opposite extreme. The single most influential actor in the conception and implementation of the *Cité* project was not a public actor from within the core executive, but a single cultural professional, Pierre Boulez; indeed, Boulez could be said without exaggeration to have 'dominated' his public interlocutors at many crucial stages of the project. A similar analysis can be applied to the contemporary music sector – in which, again, public officials followed an agenda set by a single powerful professional. At the opposite end of the continuum, we find an area of public intervention – popular music – which exemplifies the 'statist network' ideal type: fundamental policy options in the area of rock and *variétés* were set exclusively by a small and cohesive set of public actors. The 'network' tag remains appropriate, given the integration of societal interests – in this case, Bruno Lion and the *Réseau Rocks* – into the process of policy formulation. Crucially, however, the extra-governmental network members were subordinate to, and dependent on, their public interlocutors. Between the two extremes represented by the 'Boulezian' professional networks on the one hand, and the rock and *variétés* network, dominated by the Minister and his *cabinet*, on the other, the four remaining sectors exhibit hybrid characteristics. The film and theatre policy communities are situated towards the 'professional network' end of the continuum, though

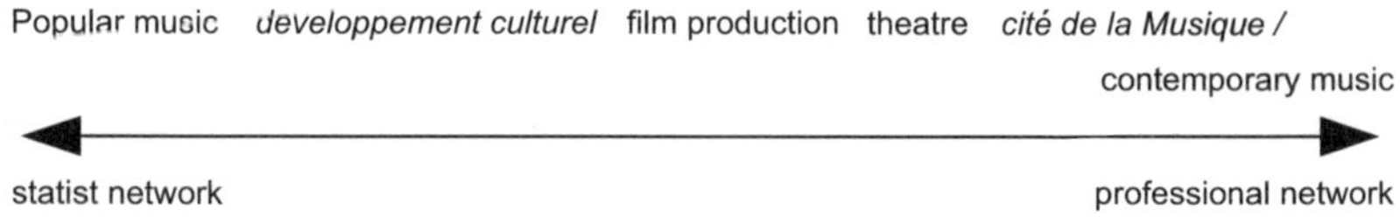

Figure 9.1 'Statist networks' and 'professional networks' in French cultural policy-making: a continuum

characterized by a much greater balance between public and private actors; the sector of *développement culturel*, dominated by the Minister and his *cabinet*, conforms more closely to the 'statist network' type (Figure 9.1).

Of course, the continuum is intended only to establish a broad pattern: it should not be taken to imply that the state could necessarily be seen as a unitary actor – especially in the case of *développement culturel*, to which we shall return at length, the core executive itself was riven by serious divisions. Furthermore, the continuum does not establish that networks categorized as 'professional' were impervious to other influences, or that societal actors had no opportunities of influencing policy in networks marked out as 'statist': clearly, even sectors otherwise characterized by professional dominance, or by a balance between professionals and public actors, were subject to occasional incursions from actors not integrated into the network. In the theatre sector, we noted occasional hierarchical interventions by the President of the Republic, which explicitly disregarded the policy solutions preferred by members of the theatre policy community. In the case of the *Cité de la Musique*, we similarly found that the President reserved for himself the right to make occasional policy choices; indeed, there were few sectors in which François Mitterrand did not on occasion intervene, usually in order to impose the appointment of a presidential *protégé* to a senior administrative or, more rarely, artistic position. Moreover, the President clearly played a decisive role in supporting the budgetary claims of the Ministry of Culture against successive Prime Ministers and Ministers of Finance.[1] Such interventions need to be acknowledged – yet they did not affect the Ministry's fundamental policy options, or perturb the functioning and composition of the core sectoral policy networks: in other words, they amounted to occasional uses of presidential power, and not to 'presidential government', or to the assertion of a presidential prerogative over a 'presidential' segment of the cultural policy area.

If we disregard such occasional exogenous interventions, a clear pattern emerges from the ordering of networks according to a 'statist network' – 'professional network' dimension. To the extent that it is possible to extrapolate from the sample of sectors considered in this book, the continuum suggests a remarkable bias of networks towards 'professional' interests. Only two sectors can be characterized as tending towards the 'statist network' ideal type; two sectors strongly exemplify professional dominance in policy-making, while two sectors are more or less balanced between the two. An even less ambiguous result emerges if we 'factor in' a quantitative weighting, and consider the distribution of budgetary allocations between the sectors concerned in the 1981–93 period. Thus, the two sectors dominated by public actors were also, among the sectors we chose to study, relatively the least advantaged at the budgetary level. The funds allocated by the Ministry of Culture to popular music – which totalled around 40 million francs in 1990 – were insignificant when compared with the subsidies of established revenue-funded institutions. In the case of *développement culturel*, funding was raised to a substantial level in the 1982 budget – but subsequently, the *direction du Développement culturel* (DDC) was subjected to greater budgetary fluctuation and rigour than any other policy sector. The DDC controlled 12.1 per cent of the Ministry's budget in 1982; by 1991, the proportion allocated to *développement culturel* had declined to 6.2 per cent. By contrast, funding levels remained virtually constant, and incomparably higher, in the 'traditional' performing arts sectors: budgetary allocations to the theatre sector thus amounted to 12.6 per cent of the Ministry's overall budget after the 1982 budget had taken effect, and stood at 12.8 per cent in 1991. The other sectors in which we found strongly 'professional' policy networks were similarly characterized by budgetary stability throughout the 1981–93 period.

In short, those sectors most representative, among our case studies, of the historical 'core' of French cultural policy were associated with a highly professionalized style of policy-making: public policy-makers acted in close, consensual collaboration with cultural professionals, and in some cases deferred entirely to the judgement of certain professionals on important policy matters. By contrast, strongly hierarchical patterns of policy-making, in which the policy agenda was determined largely or exclusively by public officials, were found only in the most 'marginal' of the areas of state intervention which we considered. In view of the dominant role which the formal political and administrative structure of the French arts funding system accords public actors, this is a striking conclusion. However, it raises two further questions. The first concerns the nature of professional influence in the policy process. We have established that professionals played a decisive role in many policy sectors: we have yet to establish which professionals, or which categories of cultural professionals, are most likely to significantly affect policy outcomes. The second question follows on from the first: what are the foundations of professional influence in the policy process, and, by implication, how do we account for variations in professional influence across sectors? As we shall see, answering the first question will go some way towards suggesting an answer to the second. We begin, then, by further refining the account of professional influence begun in this chapter so as to provide a clearer picture of the types of interest which are most likely to play a strong role in the policy-making process.

STRONG PROFESSIONALS AND WEAK PROFESSIONALS: 'INSIDERS' AND 'OUTSIDERS' IN CULTURAL POLICY-MAKING

The theme of professional influence which pervaded our sectoral case studies has been consistently interwoven with a second theme: some professionals are clearly more successful than others in affecting the policy agenda, and the inclusion of certain professionals or groups of professionals in highly integrated sectoral policy networks is accompanied, in the same sectors, by the exclusion of others. In each of our sectoral case studies, we sought to capture these distinctions between 'insiders' and 'outsiders' in the policy process by applying the typology of policy networks elaborated by Rhodes and Marsh (1992). If we compare the results from our case studies, a clear pattern emerges. Thus, in the 'core' areas of public funding – the traditional performing arts sectors and the area of film production – we find two distinct policy networks corresponding to the two endpoints on the continuum proposed by Rhodes and Marsh: on the one hand, a small, exclusive and cohesive 'policy community', and a larger, more diffuse 'issue network', on the other (Figure 9.2).

In itself, this symmetry could be no more than coincidental. Yet the relevance of the distinction between policy communities and issue networks emerges more clearly if we also take account of policy outcomes in the sectors concerned – or, more specifically, of the relative stability, or fragility, of policy outputs over the period studied. At a

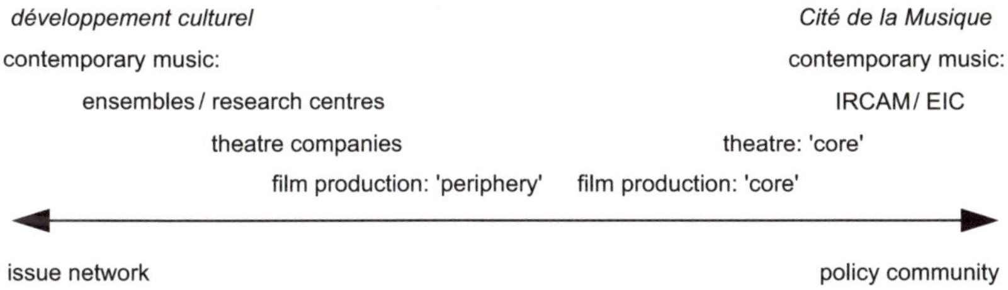

Figure 9.2. Policy communities and issue networks in French cultural policy-making: a continuum

very general level, the likelihood that certain types of network will be correlated with certain kinds of policy output is underlined by Rhodes and Marsh (1992, pp. 197–8), who note that 'the existence of a policy network or community acts as a major constraint upon the degree of policy change'. Our sectoral findings confirm this view. In the sectors we studied, the individuals and institutions clustered around the 'core' of their respective sectoral policy-making systems were most likely to benefit from a stable sectoral budget and continuity in the underlying structures and mechanisms of the sectoral funding system. They were highly resistant, moreover, to the rare attempts by actors not fully integrated into the policy community to impose policy change. As Daugbjerg (1996, p. 13) notes, '[o]utsiders' opportunities for forcing new policy roles into an old regulated sector in which a policy community exists are limited', a point which we found confirmed, notably, by the futile attempts of senior civil servants to break up the closed 'Boulezian' policy community in the contemporary music sector. More 'peripheral' institutions, by contrast, were subject to less predictable policy patterns: the budgetary and institutional fluctuation to which the smaller ensembles and research centres were subject in the contemporary music sector was a notable case in point.

Of course, both the structure of the two contrasting network types and the nature of the link between network structures and policy outcomes varied from sector to sector. Not all areas of policy-making presented as close a fit with the predictions made by Rhodes and Marsh as the contemporary music sector. In the theatre sector, we thus found that the more 'peripheral' institutions, the independent companies, though excluded from the central policy community, nonetheless benefited from a high degree of policy stability throughout the 1981–93 period, in stark contrast with the comparable 'peripheral' segment of the contemporary music sector. The difference is explained, above all, by the different strategies employed by the two groups: an aggressive 'protest politics' stance in the case of the theatre companies, facilitated by an ideological disposition to collective action; a fragmented and passive *milieu* in the case of the contemporary music sector. However, even the theatre 'issue network' remained quite distinct from the prevailing policy community: political and administrative challenges to the funding status quo were a constant possibility, and on one occasion led to overt conflict between the Ministry and the companies. The relationship between policy-makers and their 'clients' was based, in this case, not on consensus, but on a mutual 'blockage'. Again, this finding bears out the link between network structures and the susceptibility of a sector or sub-sector to policy change.

Networks and Resources: the Foundations of Professional Influence

The differentiation of sectoral policy-making systems into policy communities and issue networks raises two fundamental and related questions – first, how do we account for the fact that certain groups of professionals are tightly integrated into the policy process while others are excluded, and second, why are some cultural professionals able to penetrate the policy-making process so extensively? As we noted above, the strategies adopted by different groups of professionals do have important implications in some sectors for their ability to influence the policy process. Hence, a sustained stance of collective 'protest' action gave the independent theatre companies a degree of leverage on the Ministry of Culture which the fragmented contemporary music *milieu* was never able to achieve. However, a more comprehensive explanation of professional influence, and of the differences between network structures across sectors, must cover not only the strategies of particular categories of actors, but also, above all, their resources. We need to look, in other words, at the bases of the 'power dependence' relationship which Rhodes (1981) described as central to the emergence and persistence of cohesive policy networks.

Cultural producers differ, of course, from other groups of producers in the types of resource they are likely to have at their disposal. Rhodes and Marsh (1992, p. 199) note that 'there are two key resources which give groups privileged access to decision-making: economic position and knowledge', and suggest that these two resources can be expected to further the inclusion of the groups concerned in cohesive policy communities. The resource of 'knowledge' clearly applies as strongly in the domain of cultural production as in other policy areas: indeed, it could be argued that the importance of knowledge, or expertise, as a resource on which cultural professionals can draw in their dealings with public funding agencies is rendered all the more central by the exclusivity of the expertise offered. While policy-makers in 'conventional' areas of policy-making may be able to draw on a range of sources of technical expertise, artistic expertise is highly individualized. A certain kind of expertise – that is, the ability to produce a certain kind of cultural good – may be unique to a very small number of producers. Thus, to take the example of film production, the ability to produce or direct a commercially viable feature film may have been restricted in the period studied to a few dozen individuals in France. Not surprisingly, these were important interlocutors for the Minister of Culture, his advisers and officials at the *Centre national de la cinématographie*. In the area of contemporary music policy, to take an even more outstanding example, a certain kind of professional expertise was still more narrowly concentrated, to the point of being associated by public policy-makers with a single individual, Pierre Boulez: concomitantly, Boulez was accorded a uniquely privileged place in the policy-making process.

However, professional expertise is only one variable in the equation determining professional influence. Crucially, in order to become effective, the ability of professionals to deliver certain goods has to coincide with a perception on the part of public policy-makers that the goods concerned are necessary or, at any rate, desirable. Specifically, professional expertise can only translate into professional influence in sectors deemed of 'strategic' importance to the pursuit of wider national policy objectives – a point that has frequently been raised in connection with interest group politics in other areas of French public policy-making, such as industrial, and notably steel, policy. Both in the film sector and in the case of Boulez, the 'strategic' condition clearly applies: the protection of the French film industry has been considered a strategic objective of French policy-makers since the days of the Vichy régime; in the case of Boulez, policy-makers from Pompidou onwards have judged, rightly or wrongly, that the composer's presence and activities in France are an important source of international prestige. By contrast, in sectors or sub-sectors which are not deemed to possess a similar degree of national strategic importance, professional expertise is not as readily convertible into a power resource. The marginal position of the smaller, more marginal research centres and ensembles in the contemporary music sector is an obvious case in point, as is the sector of *développement culturel* – both of which lacked strategic importance in the eyes of policy-makers.

Professional knowledge or expertise, in short, is a decisive factor in determining professional influence, and, in the longer term, in shaping and differentiating sectoral policy networks. As for the second type of resource identified by Rhodes and Marsh, the application to cultural policy is more tenuous. 'Economic position' was not, in general, available as a resource to the groups of cultural producers we encountered in the course of our case studies. With the notable exception of the three large integrated groups in the film production sector, which had very close links with some of the largest French multinational corporations, the 'clients' of the Ministry of Culture lacked independent economic resources – not surprisingly, given their absolute dependence on the Ministry for subsidy.

However, many of the 'client' groups concerned did possess a type of resource which is virtually unique to the cultural policy area, and which we might term 'prestige' – in other words, public recognition, and a consequent ability to impact in some way on public opinion and, indirectly, on the perceived legitimacy of elected officials. In some sectors and sub-sectors, professional prestige coincided with 'strategic' professional expertise in the sense defined earlier: Boulez, again, is an obvious example. Yet in other sectors, professional prestige was sufficient by itself to affect the structure of the policy-making system. The most unambiguous example, here, is drawn from the theatre sector. The subsidized theatre is arguably, in stark contrast with the film sector, devoid of national strategic importance, given its almost exclusively domestic audience and lack of international competition. Nevertheless, leading theatre professionals occupied a position in the theatre policy community comparable to that of leading film producers in the film policy community. Thus, the influence of professional interests in the policy process was a function, in the case of the theatre sector, not so much of professional expertise as of professional prestige – which, moreover, was greatly magnified by an effective collective action strategy articulated through the theatre directors' federation, the SYNDEAC. Alongside 'knowledge', 'prestige' could thus be substituted for 'economic position' as one of two key resources which can be deployed by professionals *vis-à-vis* their governmental interlocutors. The effects of prestige could, moreover, be seen as 'self-reinforcing': in other words, as Urfalino (1989, p. 101) has pointed out, the institutional status of particular professionals within the public funding system contributes by itself to consolidating artistic reputations. In turn, this continual reinforcement of professional prestige could be said to exert a stabilizing effect on the influence of 'established' professionals in sectoral policy networks.

The account of professionals' power resources which we have set out suggests a cogent explanation both of the existence of stable network-like relationships between public and private actors, and of the differentials between different groups of professionals within and across sectors. However, our discussion of relations between public policy-makers and their professional 'clients' leads on to a wider complex of issues which we have yet to consider – that is, the linkages between networks across sectoral boundaries, and the relationship between sectoral 'cultural policies' and what we might term 'cultural policy proper'. So far, we have considered the policy-making process largely from a sectoral perspective. In fact, as we shall see, the overall policy priorities of the Ministry of Culture were in large part a function of sectoral decision-making processes: thus, professional influence, and the structure of sectoral policy networks, contributed to shaping not only sectoral outputs, but also, crucially, the 'aggregate' policy of the Ministry of Culture.

Networks, Sectors and Outcomes: Towards a General Account of Cultural Policy-making

As we noted earlier, the typology of policy networks proposed by Rhodes and Marsh provides a useful matrix for gauging and comparing the role played by different interests within particular sectors. Yet the distinction between policy communities and issue networks can be applied meaningfully not only within sectors – but also at the level of cross-sectoral comparisons.[2] In other words, sectors in which policy is made by cohesive policy communities can be expected, on the basis of the regularities posited by Rhodes and Marsh, to exhibit rather different patterns of policy from sectors in which interests are dispersed and actors are unable to agree about policy objectives.

A particularly stark contrast can be discerned, in this respect, between the 'core' sectors of public funding – in which, as we noted earlier, policy-making was highly professionalized – and the sector of *développement culturel*, which was characterized, for much of the 1981–93 period, by profound conflict, both between public and societal actors and between different actors within the core executive. The contrast between network structures in the area of *développement culturel* and in the more professionalized sectors is mirrored at the level of policy outcomes. As we noted in our introductory chapter, and as our later discussion of the institutional evolution of *développement culturel* after 1981 further spelt out, no other sector was subjected to a similar degree of budgetary and institutional uncertainty and fluctuation. Yet the link between *développement culturel* and the more professionalized sectors was more than conceptual. As we noted in our case study, the institutional and budgetary 'fragility' of *développement culturel* during the Lang years was conditioned not only by the personal and ideological tensions internal to the sector, but also by the intense competition the sector faced, notably at the budgetary level, from the Ministry's traditional *directions*. The inability of the very diverse actors concerned with *développement culturel* to develop a coherent response to these external challenges thus contributed significantly to the sector's 'decline' – further underlining the importance of a network's internal cohesion as a variable affecting policy outcomes, at both the sectoral and inter-sectoral levels.

In other words, the distinction between policy 'insiders' and 'outsiders', which we found to have accounted for significant differentials in the treatment of different types of professional interest within individual sectors, is applicable in a similar form at the 'macro' level of policy. The division of the policy-making arena into policy communities and issue networks is not based, in this case, on distinctions between different groups of professionals and their diverging degrees of access to public officials. Rather, it follows a more fundamental functional distinction between 'producer' and 'consumer' interests; specifically, it exemplifies the suggestion put forward by Rhodes and Marsh (1992, p. 199) that policy communities tend to favour the interests of producer, as opposed to consumer, groups, and mirrors the findings of many of the British case studies (Marsh and Rhodes, eds, 1992) from which the Rhodes and Marsh typology was originally derived.

Yet the significance of this analysis of *développement culturel* extends beyond the contingent circumstances surrounding the troubled history of institutions like the *direction du Développement culturel* or the *Fonds d'intervention culturelle*. First, it enables us to relate our account of the policy process to the debates about different types of policy goal in the cultural policy area, notably the contrast between 'excellence' and 'democratization', touched upon in Chapter 1; and second, it enables us to relate our analysis of cultural policy-making to the more general issue of French statism and the balance between public and private interests in French public policy-making.

GOVERNMENT, PROFESSIONALS AND NETWORKS: CULTURE, DEMOCRATIZATION AND THE 'HOSTAGE STATE'

In Chapter 1, we pointed to the dichotomy between artistic 'excellence' and cultural 'democratization' as a constant dilemma facing policy-makers since the inception of the Fifth Republic. We also noted that the policy output of the Ministry of Culture in the period studied, 1981 to 1993, was marked by a degree of continuity which it is difficult to reconcile either with the ideological impetus of the early years of Socialist rule or with the more personal commitment of the Minister of Culture to breaking down distinctions between 'high' and 'low' art. Both the distribution of budgetary allocations between cultural 'producers' and policies aimed at 'widening access', and the spread of resources between 'traditional' and 'new' artforms, remained suffused with the traditional priorities of public funders. After considering the politics of relations both between different groups of cultural producers, and between producer-based sectoral networks and the proponents of *développement culturel*, we should see the paradox as less jarring. What our analysis clearly suggests is that the 'mix' between the policy goals of 'excellence' and 'democratization' is amenable only in a limited degree to acts of political 'voluntarism'. Fundamental policy choices are subject, above all, to a range of structural constraints which make it very difficult for actors inside the core executive to impose significant policy change.

Public policy-makers, the President and the Minister of Culture foremost among them, may well 'innovate' by adding new layers to the existing body of sectoral funding policies. Pompidou thus left an important legacy for subsequent policy-makers by inviting Pierre Boulez to 'return' to France, while Mitterrand and Lang left their mark on the funding system by setting a new standard for the level of budgetary resources devoted to culture and by bequeathing to their successors a financially heavy burden of new infrastructure developments in the shape of the *grands travaux*. Moreover, policy-makers are likely to find strong support for such expansionary projects among their professional 'clientèle': the role played by Pierre Boulez in the conception of the *Cité de la Musique* offers perhaps the clearest illustration. However, innovation in the sense of redistribution – either between individual sectors, or from the professionalized 'core' of cultural policy to more experimental, non-institutional areas of funding like *développement culturel* – is likely to encounter severe resistance on the part of sectoral professionals and their allies in the administration. In other words, sectoral funding policies may be liable to unlimited growth; they are also highly resistant to policy change.

The suggestion that cultural policies tend to evolve in relative immunity to hierarchical political influences is not original to this analysis. A similar observation is central to the account of arts funding policies in liberal democracies developed by Urfalino (1989), discussed in Chapter 2. Indeed, there are close parallels between the notion that the scope for policy change is constrained by the existence of sectoral policy networks, and Urfalino's contention that the evolution of cultural policies in liberal democracies has been determined by 'dynamiques endogènes'. However, in contrast with the sociological analysis developed by Urfalino, our account of French cultural policy-making has emphasized the explicitly 'political' determinants of cultural policy: the policy process, as described in our case studies, was highly conflictual, and policy outcomes closely reflected the relative influence of different categories of professionals, civil servants and elected politicians.

In short, to the extent that it is possible to generalize from the period studied, the shape of French cultural policy is the result of complex interactive processes involving a wide variety of public and private actors. Moreover, the prevailing policy-making style is more akin to what we might term a 'low' distributive politics of interest groups competing for resources, than to the 'high' politics of sovereign presidential, or even ministerial, power – a finding which diverges both from existing accounts of cultural policy-making, and from a more general canon of assumptions about relations between the state and societal interests in Fifth Republic France.

To conclude, then, let us briefly summarize the implications of our analysis for the wider debates in which we have situated contemporary French cultural policy-making. As already suggested, the findings presented and discussed in this chapter have a bearing on two distinct areas of research. At one level, our analysis of French cultural policy-making relates to a set of very general debates and dilemmas which mark out the area of cultural policy in liberal democracies, notably with regard to the dichotomy between 'excellence' and 'democratization' as policy goals. Much of the present chapter was devoted to identifying the 'lessons' from the French experience, and we will not reiterate these in detail. However, the one essential finding is easily stated: the choice between 'excellence' and 'democratization' is likely to be settled by factors which surpass the sovereign decision-making power of elected politicians, even in a polity like France which is supposedly marked by an exceptional degree of autonomy on the part of public policy-makers. Indeed, the French example is rendered all the more significant – and generalizable – by the unusual scope of the prerogatives with which the French state is formally endowed in the area of cultural policy. Our findings thus make for a pessimistic assessment of the likelihood that a substantial policy of 'democratization' can ever be sustained in a policy arena largely colonized and dominated by influential cultural professionals.

The second broad debate to which our findings contribute concerns the relationship between the state and society, or more specifically the executive and interest groups, in Fifth Republic France. In the Introduction to this book, we noted that cultural policy has many attributes which place it at the core of a 'symbolic' area of 'high' politics in France – notably the undiluted formal authority which the political segments of the core executive enjoy even with regard to individual artistic funding decisons; the progressive appropriation of the cultural policy domain, at least in rhetoric, by successive Presidents; and, not least, the 'nation-building' and 'regime-building' roles which have been attributed to cultural policy by policy-makers since the days of the *Ancien Régime*. Here, a conclusion with more general validity can be derived from our empirical findings. Even with regard to its 'core' competences, it is possible that the French state – an entity which we considered only in the admittedly partial, but concrete and identifiable guise of the core executive – is neither as powerful, nor as autonomous *vis-à-vis* 'particularistic' societal interests, nor as united, as the statist view of French policy-making would suggest. Indeed, our central finding is perhaps best summed up by the notion of the state being 'hostage' to its interest group 'clients' – highly circumscribed, in other words, in its freedom of action and unable even remotely to exercise its formal hierarchical attributes.

This is not to suggest that the state – or, more specifically, the Ministry of Culture – has been 'captured' by societal interests in the sectors which we studied. In the elaboration of policy, the relationship between public and private interests is characterized, rather, by a high degree of consensus between public and private actors. Indeed, the thrust of our use of the policy network framework was to suggest that public and private interests in the elaboration of policy are often perceived as indistinguishable by actors on both sides of the public-private 'divide'. However, the consensus between public policy-makers and their 'clients' is underpinned, at least in those sectors in which the professional resources of 'expertise' and 'prestige' are most pronounced, by a relation of mutual 'power dependence'. In other words, the terms of the consensus between the two categories of actors are clearly circumscribed by the resources available to each side. Above all, a major change of policy which would entail breaching the consensus is likely to be sanctioned by the professional members of a given network. In this sense, public policy-makers are hostages of their professional interlocutors as much as they are initiators and shapers of policy.

Whether a similar analysis can be applied to other areas of state intervention, notably those commonly associated, like cultural policy, with a presidentialized policy style, is, of course, open to empirical investigation. In the domain of cultural policy at least, the hierarchical institutional structures of the Fifth Republic obscure an altogether more delicately balanced relationship between the state and societal interest groups. The rhetoric of presidentialism may thus be contradicted in practice by the reality of professional influence, and a focus on the 'low' politics of state-group relations may be a promising starting point for analysis of the policy process even in areas deemed to exemplify the sovereign power of the French state.

Appendix

LIST OF INTERVIEWEES

(with positions held between 1981 and 1993)

Minister of Culture

- Jack Lang

Cabinet of the Minister of Culture [Jack Lang]

- Alain Arnaud (*conseiller technique* for music)
- Francis Beck (*directeur de l'administration générale*; *directeur de cabinet*)
- Christian Dupavillon (*chargé de mission*)
- Louis Gautier (*conseiller technique* for music and theatre)
- André Larquié (*chargé de mission*)
- Thierry Le Roy (*directeur de cabinet*; *directeur de la Musique et de la Danse*)
- Claude Mollard (*chargé de mission*)
- Jacques Renard (*directeur adjoint de cabinet*)
- Jacques Sallois (*directeur de cabinet*)

Cabinet of the President of the Republic [François Mitterrand]

- Laure Adler (*conseiller technique* for culture) [telephone interview]
- Frédérique Bredin (*chargée de mission* for culture)
- Bernard Latarjet (*conseiller technique* for culture)

Cabinet of the Prime Minister [Pierre Mauroy]

- Marie-Ange Laumonier (*conseiller technique* for culture)

Cité de la Musique

- Irène Ajer (Director, *Association 'Cité de la Musique'*)
- François Barberot (architect, Atelier Christian de Portzamparc)
- Emile Biasini (*Secrétaire d'Etat* for the *grands travaux*)
- Marc Bleuse (*Directeur de la Musique et de la Danse*; Director, CNSMP; Director, *Mission musique*)
- Michel Decoust (Executive director, *Mission musique*; *Inspecteur général*, DMD)
- Dominique Dujols (Deputy director, *Mission musique*; co-author, 'rapport Fleuret')

- Alain Durel (President, *Cité de la Musique*)
- Serge Goldberg (President, *Etablissement public du Parc de la Villette*)
- François Guiguet (*chargé de mission*, *Mission musique*)
- Jean-Pierre Guillard (Director, *Mission musique*)
- Alain Louvier (Director, CNSMP)
- Brigitte Marger (Director-general, *Cité de la Musique*)
- Charlotte Nassim (*conseiller technique*, *Mission musique*)
- Guy Reibel (artistic advisor, *Mission musique*; professor, CNSMP)
- Alain Surrans (Personal assistant to Maurice Fleuret [*directeur de la Musique et de la Danse*]; Director, *Centre national d'action musicale*)

(also under 'Contemporary music policy': Pierre Boulez [President, *Ensemble InterContemporain*])

Contemporary Music Policy

- Laurent Bayle (Director, IRCAM)
- Anne Begramian (head of section, DMD)
- Pierre Boulez (composer; President, *Ensemble InterContemporain*; Director, IRCAM)
- Charley Chetrit (head of department, DMD)
- Daniel Durney (*inspecteur principal*, DMD)
- Gilbert Hotz (head of section, DMD)
- Judith Rabant (Administrator, *Ensemble de l'Itinéraire*)
- Michèle Rollot (Administrator, *Ensemble 2e2m*)

Theatre Policy

- Robert Abirached (*conseiller technique* for theatre, *cabinet* of the Minister of Culture; *Directeur du Théâtre et des Spectacles*)
- Lucien Attoun (Director, *Théâtre Ouvert* [CDN])
- Alain Crombecque (Director, *Festival d'Automne à Paris*)
- Bernard Faivre d'Arcier (*Directeur du Théâtre et des Spectacles*; Director, *Festival d'Avignon*)
- Jean-Jacques Hocquard (Administrator, *La Parole Errante* [independent company])
- Jean Jourdheuil (director; Director, *Compagnie Jean Jourdheuil* [independent company])
- René Loyon (director; President, *Action pour le jeune théâtre*; member, SYNDEAC Council; Director, *CDN de Franche-Comté*)
- Odile Quirot (*conseiller technique* for theatre, *cabinet* of the Minister of Culture)
- Christian Schiaretti (director; President, SYNDEAC; Director, *Comédie de Reims* [CDN])
- Micheline Servin (dramatist; theatre critic, *Les Temps Modernes*)
- Jean-Pierre Thibaudat (dramatist; theatre critic, *Libération*)

Film Policy

- Alain Auclaire (Deputy director-general, CNC)
- Jean-Claude Benoist (Finance director, CNC)
- René Bonnell (Director of cinema, Canal Plus; Director-general, Studio Canal Plus)
- Denis Chateau (Director-general, cinema, Pathé)
- Yves Marmion (Secretary-general, CNC; Director of production, CNC; producer; Managing director, UGC Images)
- Marc Nicolas (*conseiller technique* for cinema, *cabinet* of the Minister of Culture)
- Christian Oddos (Administrator, *Union des Producteurs de Films*)
- Pascal Rogard (Secretary-general, *Chambre Syndicale des Producteurs et Exportateurs de Films Français*; Administrator, ARP)
- Daniel Toscan du Plantier (producer; Managing director, Gaumont; Chairman, Erato; President, Unifrance Film)
- Serge Toubiana (Editor, *Cahiers du Cinéma*; Vice-chairman, *Commission des avances sur recettes*; Chairman, *Commission d'aide sélective à la distribution*)

(also under *Développement culturel*: Dominique Wallon [Director-General, CNC])

Popular Music Policy

- Jean Carabalona (*inspecteur principal*, DMD)
- Gilles Castagnac (co-founder, *Réseau Rocks*; Secretary, *Commission Consultative pour le Rock et les Variétés*)
- François de Banes-Gardonne (head of department, DMD)
- Alex Dutilh (Director, *Centre national d'action musicale*; Director, *Studio des Variétés*)
- Bruno Lion (President, *Réseau Rocks*; President, *Centre d'information du rock*; *chargé de mission* for rock and *variétés*, Ministry of Culture)
- Pierre Mayol (*chargé de recherche*, DDC)
- Dominique Ponsard (head of section, DMD)

Développement culturel

- Isabelle Blanchard (*chargée de mission*, DDC)
- Jean-Pierre Colin (*chargé de mission*, *cabinet* of the Minister of Culture)
- Alain Lennert (deputy head of department, *direction du Théâtre et des Spectacles*)
- Jean-François Marguerin (head of department, DDC)
- Hélène Mathieu (*conseiller technique*, *cabinet* of the Minister of Culture; *Déléguée au Développement et aux Formations*)
- Marie-Claude Poncet (head of department, DDC and *direction du Théâtre et des Spectacles*)
- Michel Ricard (deputy director, DDC)
- Dominique Wallon (*Directeur du Développement culturel*)

Notes

INTRODUCTION

1. See, for example, Grémion (1976), Schmidt (1990) and Worms (1966).
2. See Abirached (1992, p. 40), Caron (1989, p. 8) and Saez (1993).
3. Ritaine (1983) situates the policy pursued by the Malraux Ministry within the overall project of national integration pursued by the Gaullist regime: 'Cultural democratization should be seen less as an example of mutual concessions between the social groups, than as an illustration of the Gaullist doctrine according to which the nation's traditions give rise to the contemporary process of renewal.' (Ritaine, 1983, p. 66) Urfalino (1993: 60) similary notes that 'the striving for national greatness was probably decisive for the creation of a Ministry of Cultural Affairs in 1959'.
4. While this book is primarily focussed on the 1981–6 and 1988–93 legislatures, the overlap, in many policy areas, with the *cohabitation* period will make some reference to policy-making during the tenure of François Léotard as Minister of Culture unavoidable (see Wangermée, 1988, p. 51).

CHAPTER 1

1. Quoted in Ory (1996, p. 10).
2. See Bloch-Lainé (1982).
3. An historical assessment of Malraux's 'democratization' effort by the Council of Europe thus noted that 'Malraux did not have a real strategy for cultural democratization; he seems to have thought that, in order to win over the non-believers and to teach them how to appreciate the value of works of arts, it was enough just to present these works to them: for Malraux, all men are, in a vague kind of way, striving for culture; and those who have not been in contact with culture because of social or geographical handicaps cannot fail to be enlightened by the obvious reality of beauty, as soon as they are given the opportunity to encounter it' (Wangermée, 1988, p. 31)
4. See Foulon (1993, p. 26).
5. See Verheuge (1988, pp. 22–6).
6. See *proposition* 98: 'The state will encourage, and finance in part, the establishment in all parts of the country of centres of artistic creation, cultural activity and cultural distribution' See also *proposition* 99, which could be read as an affirmation of a more traditional *Beaux-Arts* policy with substantially reinforced means: 'Support for film-making, music, the visual arts, theatre, literature and architecture will place the country's cultural renaissance among the Socialists' priorities' (quoted in Gournay, 1988, p. 308).

7. The Socialists' electoral platform in 1981 – both presidential and legislative – did however contain explicit pledges on specific projects, notably capital projects such as the *Cité de la Musique*, and the re-imposition of unitary pricing for books.
8. In 1978, Lang claimed that once the 'mobilizing utopias' have been defined, 'the rest will follow: nothing is easier than defining a museum policy, a policy for libraries, a cinema policy, a visual arts policy' (Bredin, 1978, quoted in Clément, 1983, p. 296)
9. The evolution of cultural 'debates', and the decline of *action culturelle* as a policy goal, are discussed in great detail in Looseley (1995). See also Caune (1992, ch.10).
10. The Ministry's *Département des Etudes et de la Prospective* defines the cultural industries as follows: 'The term "cultural industries" has been taken to denote five types of activity, ranging from the production to the sale of the following cultural products: books, records, the press, the cinema, video, as well as radio and television' (*Ministère de la Culture, Département des Etudes et de la Prospective*, 1991a, p. 11)
11. It is worth noting, however, that the conceptual blurring between private and public-sector cultural production during the Lang years did not entail a significant growth in business sponsorship of the arts: in 1985, the total volume of sponsorship was estimated by the Ministry of Culture at no more than 350 million francs (Wangermée, 1988, p. 88).
12. See Wangermée (1988, p. 67).
13. In 1984, central government spending on 'culture' (including education and heritage conservation) could be broken down as follows: Ministry of Culture, 8 billion francs; Ministry of Education, 1.9 billion francs; Ministry of Foreign Affairs, 1.1 billion francs; Universities, 657.9 million francs; *Secrétariat général du Gouvernement*, 594 million francs; Ministry of Justice, 430.5 million francs; Ministry of Town Planning and Housing, 391.6 million francs; Ministry of Defence, 226.1 million francs; Ministry of Youth and Sport, 220.3 million francs.
14. Moreover, Lang took the development of fiscal and other incentives for commercial sponsorship of the arts further than any of his conservative predecessors (see Wangermée, 1988, p. 85, and Policy Studies Institute, 1990, p. 28).
15. The increases are calculated from the Ministry's figures (*dépenses ordinaires* and *autorisations de programme*, adjusted for inflation).
16. See Baleynaud (1991, p. 170).
17. Jacques Attali records, with regard to discussions of the Defferre proposals in the *Conseil des ministres*: 'Several ministers, including Jack Lang, are opposing the proposal in order to keep the right to allocate subsidies themselves. Lang has been bombarding the President with notes for two weeks' (Attali, 1993, p. 56)
18. See Baleynaud (1991, p. 169).
19. Robert Abirached, head of the Ministry's theatre directorate from 1981 to 1988, emphasizes 'the lack of understanding, the pernickety interference, the conformist attitudes and the miserliness on the part of the *collectivités locales* that had accompanied the history of the CDN and the *maisons de la culture*... Since central government had been the inspir-

ing force behind decentralization in the theatre, its partners generally expected it to maintain its role as guardian' (Abirached, 1992, p. 74). In discussing the future of the centrally funded *centres dramatiques nationaux* (CDN), Abirached suggests that transferring responsibility for the CDN to the *collectivités locales* would entail 'transforming the theatre into an art form merely reflecting its environment, and taking only modest risks' (ibid., p. 129).

20. See Baleynaud (1991, p. 158).
21. See *Ministère de la Culture* (1991a).
22. See Wangermée (1988, p. 105).
23. Following the Defferre laws, the DRAC retained the status of central government *services d'Etat*, but could 'at the demand of the regional councils be placed at their disposal "when needed"' (Friedberg and Urfalino, 1985, p. 225).
24. See Moulinier and Rizzardo (1988).
25. See Mény (1983, p. 34).
26. See Wangermée (1988, p. 110).
27. The financial transfers resulting from these *contrats* totalled 170 million francs in 1984, 278 million francs in 1985, 293 million francs in 1986 and 266 million francs in 1987 (see Wangermée, 1988, p. 110).

CHAPTER 2

1. See Colin (1994), Finkielkraut (1987), Fumaroli (1991) and Schneider (1993).
2. See Desneux (1990) and Hunter (1990).
3. See Baumol and Bowen (1966).
4. See, for example, Léonard, ed. (1993).
5. Elgie's study (1993) of prime ministerial influence deals explicitly with the impact of societal actors on the policy process; however, these are merely factored in as exogenous variables which might have a bearing on the calculations of actors within the core executive – the policy process as such is perceived as purely intra-governmental. In his account of the 1982, 1986 and 1989 French broadcasting acts, Elgie writes that 'in addition to the intragovernmental part of the process, there was also an extragovernmental element. That is to say, the situation where extra-governmental organizations intervened to try and change the wording of a bill. These organizations usually intervened after the resolution of the intragovernmental series of meetings and, thus, their influence was an element exogenous to the governmental policy process. ... Just as above it was concluded that the influence of the Prime Minister can only be appreciated alongside the influence of the other governmental actors, so the influence of the government can only be appreciated alongside the influence of these extragovernmental organizations' (Elgie, 1993, p. 69)
6. Three other areas where 'research could broaden the scope of debate beyond existing institutionalist controversies' are listed: decisional stu-

dies, the application of coalition models, and the systematic assessment of leadership influences (Dunleavy and Rodes, 1990, p. 19).

7. See Atkinson and Coleman (1989).

CHAPTER 3

1. The architect responsible for the redevelopment of La Villette, Adrien Fainsilber, had in fact included an auditorium and a dance academy in his design, neither of which was built at La Villette (see Chaslin, 1985, p. 229).
2. Interview with author, 25 January 1996.
3. See Attali (1993, pp. 185–6).
4. See Rémy (1989, p. 23) for Angremy's own account of his *mission*.
5. Any part of the site not used by the *Etablissement Public* automatically reverted to the *Ville de Paris.*
6. See Attali (1993, p. 186).
7. Interview with author, 24 July 1995.
8. The Ministry underwent numerous name changes in the 1980s: initially called 'Ministère de l'Urbanisme et du Logement', the transport portfolio was added in 1984; under *cohabitation*, it became the 'Ministère de l'Equipement, du Logement, de l'Aménagement du Territoire et des Transports', an appellation retained, with minor variations, by subsequent Socialist governments (see Collard, 1992, p. 106).
9. Interview with author.
10. Relations between the DMD and the *mission musique* were complicated not least by the fact that Lang's *directeur de la Musique* from 1981 to 1986, Maurice Fleuret, had not been given a say in Guillard's appointment, and resented the latter's autonomy.
11. Delouvrier had been appointed by Giscard d'Estaing to head the EPPV when it was created in 1979; on attaining the statutory retirement age in 1984, he was replaced by Serge Goldberg, but was given the post of 'honorary president'.
12. Interview with author, 24 July 1995.
13. The need for spatial variability applied only to a very small contemporary repertoire – including Boulez's own piece *Répons*.
14. Raymond Gallois-Montbrun, who was replaced by Marc Bleuse in 1984.
15. The *Ensemble InterContemporain*, by contrast, was funded almost exclusively by the Ministry of Culture. In fact, the music policy of the Paris city government bore the mark, throughout the 1970s and 1980s, of the composer Marcel Landowski, Chirac's main mucisal adviser and a sworn enemy of the serialist school of which Boulez was the main French exponent. Landowski's appointment as Malraux's *directeur de la Musique* in 1966 had been the main cause of Boulez's much-publicized 'break' with the publicly-funded music establishment which lasted until the accession of the cultural 'modernist' Georges Pompidou to the presidency.
16. Interview with author.

17. By way of comparison, the total cost of the *Opéra Bastille* was projected, in 1984, at 2.17 billion francs, while only 400 million francs was initially budgeted for the *Cité*.
18. Interview with author, 5 July 1995.
19. The Ministry's *inspecteurs* serve as artistic advisers to the *directeur* of the directorate in which they are based.
20. As Bleuse recalls, his appointment was explicitly decided by Delouvrier, who, given his prestige as one of France's most respected public servants, a 'grand commis de l'Etat', enjoyed an exceptional degree of autonomy in his dealings with Lang and other ministers, and who had also played a crucial role in the appointment of Jean-Pierre Guillard. In Bleuse's words, 'So I took up my position [at the *Conservatoire*], and straightaway, Paul Delouvrier ... called me. And it was Delouvrier – no one else, not Maurice Fleuret, not the Minister – who said to me, "You're the new director of the *Conservatoire*, what are your expectations?" I spoke to him for an entire morning, and he told me, "I want you to be director of the *mission musique*, otherwise people are just going to be fooling around" It was Delouvrier who decided. He called the Minister and said, "I'm constructing a building for a school, and I want to have its director, who in addition is someone you trust, since you appointed him" ... It was settled straightaway' (interview with author, 5 July 1995).
21. Interview with author.
22. Interview with author.
23. Interview with author.
24. See Chaslin (1985, p. 238).
25. See Ministère de la Culture, de la Communication et des Grands Travaux (1991a).
26. Interview with author, 5 July 1995.
27. Interview with author, 2 June 1995.
28. Interview with author, 25 January 1996.
29. Interview with author, 25 January 1996.
30. Interview with author.
31. Pierre Boulez's recollection of his encounters with the 'mission Chevrillon' is revealing: 'What was curious when these people came to see me – he came on his own once, and another time accompanied by Saint-Pulgent – was that the first time he told me: "I don't know anything about it. But I've come because I was asked by the Minister", and I felt like telling him, "But if you don't know anything about it, turn down this mission!"' (interview with author, 24 July 1995).
32. It is worth noting that the 'mission Chevrillon' was assisted by the planners of the APUR, whose perennial hope of sabotaging the Villette project coincided with the thrust of Chevrillon's proposals (see Urfalino, 1990, p. 211).
33. Interview with author, 24 July 1995.
34. Interview with author, 5 July 1995.
35. Interview with author.
36. See Hunter (1990, pp. 254–6).

37. By his own account, Bleuse – who, as a professional musician and composer, had an unusual profile for a divisional *directeur* – was 'blackmailed' into accepting the post of *directeur de la Musique* by Léotard's *cabinet*, which threatened to renege on its support for the new *Conservatoire* in case he refused; according to another senior DMD official, 'when the Minister presented Marc Bleuse [to the DMD] in 1986, Léotard came here and said, "If I am entrusting the *direction de la Musique* to Marc Bleuse, it is because he is someone who persuaded me so effectively that La Villette had to continue that I've decided, first, that La Villette had to continue, and second, that this was the man to whom the direction de la Musique had to be entrusted"' (interview with author).
38. Interview with author.
39. Interview with author.
40. *Mission musique* official (interview with author).
41. Biasini replaced the *mission de coordination* which had followed the *grands travaux* at a purely technical level during the first *septennat* (see Collard, 1992).
42. Interview with author, 2 June 1995.
43. Boulez had known Biasini since the latter was expelled from his position at the Ministry of Culture for siding with Boulez against Malraux in 1966.
44. Interview with author, 2 June 1995.
45. Interview with author, 24 July 1995.
46. Ajer had also worked closely with Biasini in the 1960s; her appointment could thus be construed, in the words of a senior *cabinet* source, as resulting from the 'Boulez–Biasini alliance' (Interview with author).
47. Interview with author.

CHAPTER 4

1. See Wangermée (1988, p. 149).
2. For a detailed historical account of relations between Boulez and the Ministry of Culture in the 1960s and 1970s, see Born (1995, pp. 75–91).
3. See Durney (1993). Boulez was supported by Emile Biasini, then the second most senior civil servant in the Ministry for Cultural Affairs – who was consequently dismissed by Malraux. It is worth noting that Boulez and Biasini would later collaborate closely on the *Cité de la Musique* project.
4. Legally, IRCAM was given the statutes of a private association in accordance with the provisions of the 1901 law on *associations*; at the same time, it was tied contractually to the *Centre Pompidou*. Funding for IRCAM is thus formally the responsibility of the *direction de l'Administration générale* of the Ministry of Culture, which controls *établissements publics* like the *Centre Pompidou*.
5. Interview with author.
6. Interview with author.

7. Interview with author, 24 July 1995.
8. Interview with author.
9. See *Ministère de la Culture* (1985a), 'La musique et la danse', p. 13.
10. It is worth noting that, between 1978 and 1993, IRCAM's self-generated income rose from around 2.91 million francs to 14.28 million francs (*Le Monde*, 12 April 1995).
11. Interview with author.
12. Interview with author.
13. Interview with author.
14. Fleuret was initially opposed, not least, to the building of the *Opéra Bastille* – without, however, being in a position to influence policy either on the capital project as such or on revenue funding for the Paris Opera.
15. Interview with author.
16. Interview with author.
17. Interview with author.
18. Interview with author.
19. The conflict became public when Boulez published an article in *Le Monde,* dated 3 February 1990, in which he accused Schneider of 'a total lack of interest for our concerts ... which borders on malevolence'.
20. Schneider was a member of the *Cour des Comptes*.
21. In an interview with *Le Monde* (13 September 1990) in which he responded to Boulez's criticisms, Schneider noted that 'the state is not a self-service store' and, hinting at Boulez's influence over important elements of music policy, asserted: 'Is it absurd to entrust an administration to an administrator? ... I expect, in return, that the artist does not take himself for a public manager' Schneider elaborated on this perspective in the polemical essay against the policies of the Lang Ministry which he published after his resignation (Schneider, 1993).
22. Interview with author.
23. Interview with author, 24 July 1995.
24. Interview with author.
25. The only research centre receiving more substantial permanent funding, the *Groupe de recherche musical* (GRM), was attached to the *Institut National de l'Audiovisuel* (INA) and thus funded by the ministry responsible for communications, rather than the Ministry of Culture.
26. See *Ministère de la Culture* (1985a), 'La musique et la danse', p. 16. Apart from the institutions referred to, the contemporary music budget comprised support for specialized festivals, subsidies for 'musical theatre' projects, and direct patronage of composers by means of commissions and bursaries.
27. Besides his career as music critic of the *Nouvel Observateur*, Fleuret had in the 1970s been director of several contemporary music festivals.
28. Interview with author.
29. The two most prestigious and most generously funded ensembles, 2e2m and *L'Itinéraire*, were established by the composer-conductors Paul Méfano and Michael Levinas; similarly, the largest of the electro-acoustic studios supported by the Ministry were linked to composers, such as Iannis Xenakis, Pierre Henry and Jean-Claude Eloy.
30. Interview with author.

31. Interview with author.
32. Interview with author.
33. The figure is calculated from data provided by 2e2m.
34. Interview with author.
35. Interview with author.

CHAPTER 5

1. Four of the *théâtres nationaux* are located in Paris (the *Comédie Française*, the *Théâtre National de Chaillot*, the *Odéon-Théâtre de l'Europe* and the *Théâtre National de la Colline*), the *Théâtre National de Strasbourg* being the only provincial institution in this category.
2. A small number of CDN – notably those in Nanterre and Villeurbanne – are generally considered to have a status equivalent to that of the *théâtres nationaux*, reflected in subsidy levels on a par with those of the smaller *nationaux*.
3. Before assuming responsibility for theatre policy at the Ministry of Culture, first as *conseiller technique*, then as *directeur du Théâtre*, Robert Abirached had worked as a theatre critic and academic. Abirached was among a small number of close personal friends of Jack Lang's appointed to important posts in the Ministry in 1981.
4. Interview with author, 15 March 1995.
5. Interview with author, 15 March 1995.
6. Interview with author.
7. Interview with author.
8. Interview with author.
9. Interview with author.
10. Interview with author.
11. Interview with author, 15 March 1995.
12. Interview with author.
13. Interview with author.
14. The expansion of funding was accompanied by a devolution of the advisory commissions to the Ministry's *directions régionales des Affaires culturelles*.
15. Interview in *Libération*, 7 June 1989.
16. Interview with author.
17. Interview with Robert Abirached in: *Acteurs* 1988 (4).
18. Interview with author.
19. Interview with author.
20. Interview with author.
21. Interview with author.
22. Savary had previously been director of the Montpellier CDN; one consequence of his move to Chaillot, much remarked upon by his critics, was the cancellation of the large deficit he had accumulated as director in Montpellier.
23. Interview with author.
24. Interview with author, 25 January 1996.
25. Interview with author, 15 March 1995.

26. Interview with author.
27. The term was employed by several officials and theatre professionals in interviews with the author.

CHAPTER 6

1. See Hayward (1993, p. 386).
2. Technically, the *compte de soutien* is included in the annual state budget; however, all decisions pertaining to its volume and allocation are made by the Ministry of Culture.
3. Interview with author.
4. See Karmitz (1994, p. 147).
5. Interview with author, 25 January 1996.
6. See Frodon (1995, p. 649).
7. Interview with author.
8. Interview with author.
9. Interview with author.
10. The tax breaks associated with the SOFICA were exceptionally generous: individuals could deduct their investment up to a limit of 25 per cent of their total income; companies could claim tax immunity on up to 50 per cent of the sum invested.
11. See Racine (1994).
12. Interview with author.
13. Interview with author.
14. Interview with *cabinet* member.
15. Interview with author.
16. Interview with author.
17. Bredin's father, Jean-Denis Bredin, was author of the report on film policy submitted to Jack Lang on 3 November 1981. Frédérique Bredin would subsequently become François Mitterrand's cultural adviser during *cohabitation*; after Mitterrand's re-election in 1988, she became first a Socialist deputy and later Minister for Youth and Sport.
18. Interview with author.
19. Interview with author.
20. Interview with author.
21. Interview with author.
22. See Karmitz (1994).
23. Interview with *cabinet* member.
24. Interview with author.
25. Interview with author.
26. Interview with author.
27. Interview with author.
28. Interview with author.
29. Interview with author.
30. Interview with author.
31. Interview with author, 25 January 1996.

CHAPTER 7

1. See Looseley (1995, p. 131) and Dubois (1993) for an overview of cultural debates in the 1980s.
2. Speech to the National Assembly, 17 November 1981, quoted in Looseley (1995, p. 123).
3. Polemical essays by Finkielkraut (1987), Fumaroli (1991) and Schneider (1993) stand out as the most noteworthy contributions to the debate about the extension of state funding to 'low' culture.
4. The term '*variétés*' is generally taken to designate a wide range of French popular music styles not adequately described by the 'rock' label with its 'Anglo-Saxon' connotations.
5. Interview with author.
6. Interview with author.
7. Interview with author.
8. See *Ministère de la Culture* (1985a), 'La musique et la danse,' p. 26.
9. Interview with author.
10. Interview with author.
11. Interview with author.
12. Interview with author.
13. Interview with author. In 1985, Jack Lang, in response to criticism of the Ministry's progressive withdrawal from the CRC venture, effectively declared that the CRC and their survival had always been first and foremost the responsibility of the *collectivités locales* (Lang, 1985b, p. 26).
14. Interview with author.
15. Interview with author.
16. Interview with author.
17. The construction of the Paris *Zénith* – which was followed by a series of provincial *Zéniths* – was financed by the *Caisse des dépôts*, and its management was transferred to a private company. The *Zénith* had originally been intended as a temporary substitute for the permanent popular music auditorium planned at Bagnolet as part of François Mitterrand's *grands travaux* programme, which was never built.
18. Interview with author.
19. Interview with author.
20. See Teillet (1992, p. 313).
21. 'Convention relative à la création d'un réseau d'information sur les musiques "rock",' 6 March 1986.
22. Interview with author.
23. The Léotard Ministry continued to fund the CIR after 1986; similarly, after 1993, Lang's conservative successors Jacques Toubon and Philippe Douste-Blazy maintained the status of the centre and its successor, the *Centre d'Information et de Ressources pour les Musiques Actuelles*, as the Ministry's privileged interlocutor in the popular music sector.
24. Interview with author.
25. It is worth noting that, throughout the evolution of its 'rock policy,' the Ministry lacked reliable data about the rock *milieu* at which its policies

were aimed. Thus, the policy was premised on the notion that rock constituted a widespread cultural practice, and that a 'rock policy' could address itself to a vast amateur constituency. In order to provide a statistical basis for this assertion, Ministry officials frequently spoke of France having 25,000 rock bands; indeed, Jack Lang in 1985 cited the figure of '30,000 rock bands in France' (Interview with Jack Lang, *Ouest-France*, 19 March 1985). The process by which the '25,000 bands' were arrived at, as recounted by a former member of the *Réseau Rocks*, is revealing: 'In that period, all our dossiers started with: "There are 25,000 rock bands in France. Because in fact it's the official figure provided by the Ministry of Culture" ... Basically, what happens at the time is that, one day, the people from the Ile-de-France DRAC just ask Bruno Lion, who is their interlocutor in this area: "How many rock bands are there in the Ile-de-France region?" ... And Bruno, just like that, says, "5,000". And the Ile-de-France DRAC puts down, "There are 5,000 rock bands in Ile-de-France" No one has ever checked this. It goes up to the level of the Ministry, and the Ministry says, "Right, Ile-de-France represents roughly one fifth of the nation's cultural activity, so there are 25,000 rock bands in France" The CIR, which has been counting artists for 10 years, has never succeeded in getting beyond the figure of roughly 5,000 for all of France' (interview with author).

26. Quoted in Looseley (1995, p. 160).
27. See Teillet (1992, p. 390).
28. Interview with author.
29. Interview with author, 23 November 1994.
30. 'FAIR,' introductory leaflet published by FAIR (1994).
31. Interview with author.
32. See Schneider (1993).
33. Interview with author.
34. Interview with author.
35. The funds attributed specifically to rock and *variétés* in the 1990 budget, by both the *direction de la Musique* and other *directions*, totalled around 43 million francs by the Ministry's own estimates, an increase of 50 per cent over the previous year's budget (see *Ministère de la Culture* [1989c]).
36. Interview with author, 23 November 1994.
37. Interview with author.
38. Elgie, in his discussion of different varieties of 'core executive' decision-making, does not consider the possibility that policy-making styles may vary across different sectors even within a single ministry's area of responsibility; we thus use 'ministerial government' in a more discriminating fashion, to apply to a clearly delineated policy sector in a specified period.
39. Interview with author.

CHAPTER 8

1. The following characterization of the DDC by Lang's first *directeur du Théâtre*, Robert Abirached, is indicative of the tension between the new

direction's political rationale and the concerns of sectoral policy-makers: 'I'm not a member of the PS, they didn't have anything to do with the *direction du Théâtre*, and if they wanted to intervene, it was sufficient for me to say, "I don't want anyone to interfere, leave me alone", and I would be left alone. On the other hand, over there [at the DDC], it was work with the regions, with the Socialist partners, it was ideological work – it was what remained. What remained of a Socialist programme for culture' (interview with author, 15 March 1995).

2. 'Objectifs et moyens d'une Direction du Développement Culturel', internal memorandum to Lang, 8 September 1981.
3. See Wangermée (1988, p. 213).
4. A *direction*, or directorate, is the highest subdivision within a Ministry's administrative hierarchy; its head, a *directeur*, is appointed by decree in the Council of Ministers and is accountable only to the Minister.
5. Sallois and Mollard were members of the same *grand corps*, the *Cour des Comptes* – and both, ironically, had been members of the same PSU section in the 1960s.
6. Interview with author. Mollard was subsequently appointed head of the newly created *délégation aux Arts plastiques*.
7. Interview with author.
8. Interview with author. Wallon had been a left-wing student leader during the Algerian war.
9. Interview with author.
10. Interview with author, 25 January 1996.
11. Interview with author.
12. Interview with author. In the words of the same official, the DDC failed to conform to Lang's strategy: 'Lang was always surprised to find that there was no reservoir. So, he certainly managed to take a little money for SOS Racisme or things like that, but on the whole, it really was absolutely marginal.'
13. Interview with author, 25 January 1996.
14. In September 1981, Dominique Wallon had proposed that the new *direction* have 'a role of stimulation and of co-ordination – particularly with regard to everything touching on regional action – *vis-à-vis* the Ministry's other directorates' ('Objectifs et moyens d'une Direction du Développement Culturel', internal memorandum to Lang, 8 September 1981).
15. Interview with author, 15 March 1995.
16. Interview with author.
17. See Saez (1988) and Looseley (1995, pp. 106–8) for a more detailed discussion of the ideological debates surrounding policy towards the EAC.
18. Interview with author.
19. Interview with author.
20. Interview with author.
21. Interview with author.
22. Interview with author.
23. Interview with author.
24. Interview with author.

25. Interview with author, 25 January 1996.
26. Interview with author.
27. Interview with author.
28. Interview with author.
29. Interview with author, 25 January 1996.
30. Interview with author.
31. Interview with author.
32. Interview with author.
33. Interview with author.
34. Interview with author, 25 January 1996.
35. The 'inventor' of the FIC, Claude Mollard, then *rapporteur* of the Plan's cultural commission, would in 1981 join Lang's *cabinet* as *chargé de mission* with responsibility for the budget; he was himself later to criticize the FIC's abolition.
36. Interview with author.
37. Interview with author.
38. Interview with author.
39. Interview with author.
40. Interview with author.
41. Interview with author.
42. Interview with author.

CONCLUSION

1. See Cole (1994, pp. 91–2) and Colin (1994, p. 167).
2. While Rhodes and Marsh developed their typology on the basis of sectoral-level case studies, there is no reason why the concepts should not be applied at either the sub-sectoral or the super-sectoral level: as Marsh (1995, p. 142) underlines, 'we consider all these questions relating to the formation of networks, the level at which they are constituted and relations between different networks, to be empirical questions'.

Bibliography

Abirached, R. (1988) 'Où va le théâtre? Entretien avec Robert Abirached', *Esprit* 143 (October): 111–18.

Abirached, R. (1992) *Le théâtre et le prince*, Paris: Plon.

Après-demain (1990) 'La culture en question', *Après-demain* 322 (March): 2–38.

Association française de science politique, ed. (1986) *L'Etat devant les cultures régionales et communautaires. Colloque, 23–25 janvier 1986*, Paris: AFSP.

Atkinson, M.M. and Coleman, W.D. (1989) 'Strong States and Weak States: Sectoral Policy Networks in Advanced Capitalist Economies', *British Journal of Political Science* 19: 47–67.

Attali, J. (1993) *Verbatim (I)*, Paris: Fayard.

Attali, J. (1995) *Verbatim (III)*, Paris: Fayard.

Baleynaud, P. (1991) 'La Culture: l'oubliée de la décentralisation?' *Revue du Droit Public et de la Science Politique en France et à l'Etranger* (January/February): 147–95.

Bauer, M. and Cohen, E. (1981) *Qui gouverne les groupes industriels?*, Paris: Seuil.

Baumol, W.J. and Bowen, W.G. (1966) *The Performing Arts: the Economic Dilemma*, New York: Twentieth Century Fund.

Beaulac, M. and Grandmont, G., eds. (1992) *Pouvoirs politiques et politiques culturelles: enjeux nationaux*, Montreal: HEC.

Becker, H. (1982) *Art Worlds*, Berkeley: University of California Press.

Benson, J.K. (1982) 'Networks and Policy Sectors: A Framework for Extending Interorganizational Analysis', in D. Rogers and D. Whitten, eds., *Interorganizational Coordination*, Ames, Iowa: Iowa University Press.

Benz, A., Scharpf, F.W. and Zintl, R. (1992) *Horizontale Politikverflechtung: zur Theorie von Verhandlungssystemen*, Frankfurt: Campus.

Benz, A. (1993) 'Commentary on O'Toole and Scharpf: The Network Concept as a Theoretical Approach', in Scharpf, ed. (1993): 167–75.

Biasini, E., et al. (1989) *Le Grand Louvre*, Paris: Electra Moniteur.

Biasini, E. (1995) *Grands Travaux*, Paris: Odile Jacob.

Birnbaum, P. (1985) *Les élites socialistes au pouvoir: les dirigeants socialistes face à l'Etat, 1981–1985*, Paris: PUF.

Blaug, M., ed. (1976) *The Economics of the Arts*, Boulder: Westview Press.

Bloch-Lainé, F. (1982) *La France en Mai 1981*, Paris: La Documentation française.

Bloom, Alan (1993) *L'Ame désarmée – Essai sur le déclin de la culture générale*, Paris: Julliard.

Bonnell, R. (1989) *La vingt-cinquième image: une économie de l'audiovisuel*, Paris: Gallimard.

Bonnier, H. (1992) *Lettre recommandée à Jack Lang et aux fossoyeurs de la culture*, Paris: Rocher.

Born, G. (1995) *Rationalizing Culture: IRCAM, Boulez, and the Institutionalization of the Musical Avant-Garde*, Berkeley: University of California Press.

Bosquet. A. (1981) *Lettre au Président de la République sur les insuffisances de notre culture*, Paris: Luneau-Ascot.
Boublil, A. (1991) *Le pouvoir de l'art*, Paris: Belford.
Boulez, P. (1988) 'Du Domaine musical à l'IRCAM. Entretien avec P.-M. Menger', *Le Débat* 50 (May/August): 257–266.
Bourdieu, P. (1979) *La Distinction*, Paris: Minuit.
Bredin, J.-D. and Lang, J. (1978) *Eclats*, Paris: Jean-Claude Simoen.
Bredin, J.-D. (1981) *Rapport de la Mission de réflexion et de proposition sur le cinéma. Rapport remis à M. le Ministre de la Culture.* Paris: Ministère de la Culture.
Broussky, F. (1991) *Démocratisation culturelle ou démocratie culturelle? L'apport des penseurs de l'Ecole de Francfort*, *mémoire* (unpublished), Paris: Institut d'Etudes Politiques.
Busson, A. and Evrard, Y. (1987) *Portrait économique de la culture*, Paris: La Documentation française.
Cabanne, P. (1981) *Le pouvoir culturel sous la Ve République*, Paris: Orban.
Caron, R. (1989) *L'Etat et la culture,* Paris: Economica.
Caune, J. (1981) *Esthétique de l'animation culturelle: pour un autre statut du processus artistique*, Grenoble: Centre de recherche sur l'édition.
Caune, J. (1992) *La culture en action: de Vilar à Lang le sens perdu*, Grenoble: Presses Universitaires de Grenoble.
Cavanagh, M. (1996) *Policy Networks, Policy Outcomes and Offshore Health and Safety in the Oil and Gas Industries of Britain and Norway*, Paper to the Political Studies Association Annual Conference (Panel: Comparing Policy Networks), University of Glasgow, 10 April 1996.
Centre d'information du rock et des variétés (1989) *L'Officiel du Rock 90*, Paris: CIR.
Centre d'information du rock et des variétés (1994) *L'Officiel 94: Guide annuaire des musiques actuelles*, Paris: CIR.
Chalendar, J. de and Brébisson, G. de (1987) *Mécénat en Europe,* Paris: La Documentation française.
Chambart de Laurre, P.-H. (1975) *La culture et le pouvoir*, Paris: Stock.
Chaslin, F. (1985) *Les Paris de François Mitterrand: histoire des grands projets architecturaux*, Paris: Gallimard.
Chazel, F. (1987) *Pratiques culturelles et politiques de la culture*, Talence: Maison des sciences de l'homme de l'Aquitaine.
Chougnet, J.-F. (1993) 'Le financement de la culture en France: alchimie ou anarchie?', in Léonard, ed. (1993): 56–63.
Clément, C. (1982) *Rêver chacun pour l'autre,* Paris: Fayard.
Cluzel, J. (1984) *Les pouvoirs publics et la transmission de la culture*, Paris: LGDJ.
Cohen, E. (1989) *L'Etat brancardier*, Paris: Calmann-Lévy.
Colas, D., ed. (1988) *L'Etat et les corporatismes: travaux de la Mission sur la modernisation de l'Etat*, Paris: PUF.
Cole, A. (1994) *François Mitterrand: a study in political leadership*, London: Routledge.
Colin, J.-P. (1986) *La Beauté du manchot: culture et différence*, Paris: Publisud.
Colin, J.-P. (1990) 'La Culture et son administration', *Commentaire* 50: 344–46.
Colin, J.-P. (1994) *L'acteur et le roi*, Chêne-Bourg: Georg.

Colin, J.-P. and Séloron, F. (1994) *Le mandarin étranglé, réflexion sur la fonction sociale de l'art*, Paris: Publisud.

Collard, S. (1992) 'Mission impossible: les chantiers du Président', *French Cultural Studies* ii (June): 97–132.

Collard, S. (1996) 'Politics, culture and urban transformation in Jacques Chirac's Paris 1977–1995', *French Cultural Studies* vii (February): 1–31.

Commissariat général du Plan (1993) *La création face aux systèmes de diffusion. Rapport du groupe "création culturelle, compétitivité et cohésion sociale". Présidé par Marin Karmitz*, Paris: La Documentation française.

Conseil de l'Europe, Conseil de la Coopération Culturelle, Programme Européen d'Evaluation, ed. (1988) *La politique culturelle de la France*, Paris: La Documentation française.

Crane, D. (1992) *The Production of Culture*, Newbury Park, CA.: Sage.

Crépu, M. & Lenglart, D. (1985) 'La république des architectes', *Esprit* 109 (December): 72–5.

Cummings, M.C. and Katz, R.S., eds. (1987) *The Patron State*, New York: Oxford University Press.

Cummings, M.C. and Schuster, J.M.D. (1989) *Who's to Pay for the Arts?*, New York: ACA Books.

Dagnaud, M. and Mehl, D. (1982, 1988) *L'élite rose*, Paris: Ramsay.

D'Angelo, M. (1983) *La Muse et la cité: l'institution musicale et le système social français, Thèse 3e cycle, sociologie* (unpublished), Paris: Institut d'Etudes Politiques.

Daugbjerg, C. (1996) *Policy Networks and Agricultural Policy Reforms in Sweden and the European Community: Explaining Different Policy Outcomes*, Paper to the Political Studies Association Annual Conference (Panel: Comparing Policy Networks), University of Glasgow, 10 April 1996.

Debray, R. (1979) *Le pouvoir intellectuel en France*, Paris: Ramsay.

Délégation à l'action territoriale et à l'aménagement du territoire (1992) *L'aménagement culturel du territoire*, Paris: La Documentation française.

Desneux, R. (1990) *Jack Lang: la culture en mouvement*. Paris: Favre.

DiMaggio, P.J. (1981) *The Impact of Public Funding on Organizations in the Arts*, Yale Program on Non-Profit Organizations Working Paper 31, New Haven: Yale University.

DiMaggio, P.J. (1986a) 'Introduction', in DiMaggio, ed. (1986): 3–13.

DiMaggio, P.J. (1986b) 'Can Culture Survive the Marketplace?', in DiMaggio, ed. (1986): 65–92.

DiMaggio, P.J. (1986c) 'Cultural Entrepreneurship in Nineteenth-Century Boston', in DiMaggio, ed. (1986): 41–61.

DiMaggio, P.J. and Powell, W.W. (1983) 'The Iron Cage Revisited: Institutional Isomorphism and Collective Rationality in Organizational Fields', *American Sociological Review* 48 (1983): 147–60.

DiMaggio, P.J. and Useem, M. (1978a) 'Cultural Property and Public Policy: Emerging Tensions in Government Support for the Arts', *Social Research* 45 (2): 356–89.

DiMaggio, P.J. and Useem, M. (1978b) 'Social Class and Arts Consumption', *Theory and Society* 5: 141–61.

DiMaggio, P.J., ed. (1986) *Nonprofit Enterprise in the Arts*, New York: Oxford University Press.

Djian, J.-M. (1982) *Le Fonds d'intervention culturelle ou l'instrument d'une politique culturelle de l'Etat en matière d'innovation, thèse de troisième cycle, science politique* (unpublished), Dijon: Université de Dijon.

Donnat, O. (1988) 'Politique culturelle et débat sur la culture', *Esprit* 144 (November): 90–101.

Donnat, O. (1991) 'Démocratisation culturelle: la fin d'un mythe', *Esprit* 170 (March/April): 65–82.

Donnat, O. and Cogneau, D. (1990) *Les pratiques culturelles des Français 1973–1989,* Paris: La Découverte/La Documentation française.

Dowding, K. (1991) *Rational Choice and Political Power*, Aldershot: Edward Elgar.

Dowding, K. (1994) 'Policy Networks: Don't Stretch a Good Idea too Far', in Dunleavy, P. and Stanyer, J., eds. *Contemporary Political Studies, 1994, vol. I*, Belfast: The Political Studies Association of the United Kingdom, 59–78.

Dowding, K. (1995) 'Model or Metaphor? A Critical Review of the Policy Network Approach', *Political Studies* 43 (1), 136–58.

Dubois, V. (1993) 'Politiques culturelles et polémiques médiatiques', *Politix* 24 (December): 5–19.

Dumazedier, J. and Girard, A. (1993) 'Sciences sociales et politiques culturelles. Entretiens avec Joffre Dumazedier et Augustin Girard', *Politix* 24 (December): 57–77.

Dunleavy, P. (1991) *Democracy, Bureaucracy and Public Choice*, London: Harvester.

Dunleavy, P. and O'Leary, B. (1987) *Theories of the State*, Basingstoke: Macmillan.

Dunleavy, P. and Rhodes, R.A.W. (1990) 'Core Executive Studies in Britain', *Public Administration* 68 (Spring): 3–28.

Dupavillon, C. (1989) *Le Triomphe des Arcs*, Paris: Gallimard.

Dupuis, X. (1981) *Essai sur les pratiques culturelles de l'Etat: l'exemple de la musique, thèse de troisième cycle* (unpublished), Paris: Université Paris XIII.

Dupuis, X. (1983) 'La surqualité: le spectacle subventionné malade de la bureaucratie?', *Revue économique*, 34 (6): 1089–115.

Dupuis, X. (1985) 'La Microéconomie du spectacle vivant', in Service des études et recherches and Association pour le développement et la diffusion de l'économie de la culture, eds. *L'Economie du spectacle vivant et l'audiovisuel*, Paris: La Documentation française, 71–99.

Durney, D. (1992) *1989 ou le déclin de la République musicale* (unpublished typescript).

Durney, D. (1993) 'The State, the Creator and the Public. Twenty Years of Policy-making for Musical Creation: France 1966–86', *Contemporary Music Review* 5: 1–16.

Duyamet, P., ed. (1982) *L'Impératif Culturel. Rapport du Groupe de Travail Long Terme Culture*, Paris: La Documentation française.

Elgie, R. (1992) *The Influence of the French Prime Minister in the Policy-making Process, 1981–1991*, PhD thesis, London: University of London.

Elgie, R. (1993) *The Role of The Prime Minister in France*, Basingstoke: Macmillan.

Eling, K. (1996) 'Presidentialism and the *grands travaux*: the case of the Cité de la Musique', *Modern & Contemporary France* NS 4 (3): 331–340.

Emmanuel, P. (1980) *Culture, noblesse du monde: histoire d'une politique*, Paris: Stock.
Enthoven, J.-P. (1983) 'Jack Lang comme si vous y étiez', *Le Nouvel Observateur* 994 (1 December 1983): 114–115.
Esprit, ed. (1984) 'Mainmise sur la culture', *Esprit* 3 (March): 45–110.
Farchy, J. and Sagot-Duvauroux, D. (1994) *Economie des politiques culturelles*, Paris: PUF.
Feigenbaum, H. (1985) *The Politics of Public Enterprise*, Princeton: Princeton University Press.
Finkielkraut, A. (1987) *La Défaite de la Pensée,* Paris: Gallimard.
Forbes, J. (1987) 'Cultural Policy: the Soul of Man under Socialism', in Mazey, S. and Newman, M., eds. *Mitterrand's France*, London: Croom Helm: 131–65.
Forest, P. (1993) 'Le concept contemporain de culture', in Léonard, ed. (1993): 3–11.
Fosseyeux, J. (1991) *Les relations du ministère de la culture et de ses usagers* (unpublished), Paris: Ministère de la Culture, Direction de l'Administration générale.
Foulon, C.-L. (1990) 'Des Beaux-Arts aux affaires culturelles (1959–1969): les entourages d'André Malraux et les structures du Ministère', *Vingtième siècle* 28 (December): 29–40.
Foulon, C.-L. (1993) 'L'Etat et le gouvernement des Affaires culturelles de 1959 à 1974', in Léonard, ed. (1993): 18–33.
Four, P.-A. (1988) *L'action culturelle dans le secteur des arts plastiques: la mise en place des FRAC, 1981–1985, Mémoire DEA, Etudes politiques* (unpublished), Paris: Institut d'Etudes Politiques.
Four, P.-A. (1993) 'La compétence contre la démocratisation? Création et re-création des Fonds régionaux d'art contemporain', *Politix* 24 (4): 95–114.
Friedberg, E. and Urfalino, P. (1983) *Le jeu du catalogue: les contraintes de l'action culturelle dans les villes*, Paris: La Documentation française.
Friedberg, E. and Urfalino, P. (1984a) *La décentralisation culturelle: la culture au service des régions*, Paris: Ministère de la Culture, Direction du Développement culturel.
Friedberg, E. and Urfalino, P. (1984b) 'La gestion des politiques culturelles municipales: de l'inflation culturelle à l'évaluation politique', *Politique & Management Public*, 2 (1): 3–26.
Friedberg, E. and Urfalino, P. (1985) 'La décentralisation culturelle: l'émergence de nouveaux acteurs', *Politique & Management Public* 3 (2): 215–26.
Frodon, J.-M. (1995) *L'âge moderne du cinéma français*, Paris: Flammarion
Fumaroli, M. (1982) 'De Malraux à Lang: l'excroissance des Affaires culturelles', *Commentaire* 5 (18): 244–59.
Fumaroli, M. (1990) 'La culture et les loisirs: une nouvelle religion d'Etat', *Commentaire* 13 (51): 425–35.
Fumaroli, M. (1991) *L'Etat Culturel*, Paris: De Fallois.
Fumaroli, M. (1992) 'L'Etat, la culture et l'esprit: questions à Marc Fumaroli', *Le Débat* 70: 76–83.
Fumaroli, M.(1993) 'Pour une rupture', *Le Monde*, 25 March 1993.
Gain, N. (1974) *La neutralité de l'Etat en matière culturelle, Thèse de doctorat* (unpublished), Paris: Universitè Paris II.

Gilbert, C. and Saez, G. (1982) *L'Etat sans qualités,* Paris: PUF.

Giordan, H. (1982) *Démocratie culturelle et droit à la différence. Rapport présenté à Jack Lang, Ministre de la Culture*, Paris: La Documentation française.

Girard, A. (1981) 'Policy and the Arts: the Forgotten Cultural Industries', *Journal of Cultural Economics* 5: 61–8.

Girard, A. (1982) *Développement culturel: expérience et politique*, Paris: Dalloz-UNESCO.

Girard, A., ed. (1982) *Pratiques culturelles des Français. Description socio-démographique. Evolution 1973–81*, Paris: Dalloz.

Girard, A. (1986) 'L'enjeu culturel', in A. de Tarlé., ed., *Les Enjeux de la fin du siècle*, Paris: Desclée de Brouwer, 69–93.

Glayman, C. (1993) 'Repères sur la musique contemporaine', *Esprit* 193 (July): 123–39.

Gollier, S. (1977) *Politique culturelle et budget culturel, mémoire DEA*, (unpublished), Paris: Université Paris I.

Gournay, B. (1988) 'Rapport National', in Conseil de l'Europe, Conseil de la coopération culturelle, Programme Européen d'Evaluation, ed. (1988): 275–394.

Grémion, P. (1976) *Le pouvoir périphérique*, Paris: Seuil.

Griotteray, A. (1983) 'Pour une politique des beaux-arts', *Revue des deux mondes* 12 (December): 557–66.

Guilloux, J. (1990) 'L'Opéra dans tous ses états', *Revue politique et parlementaire* 92 (946).

Guy, M. (1988) 'L'image et l'écrit', *Esprit* 145: 106–12.

Haas, P. (1989) 'Do Regimes Matter? Epistemic Communities and Mediterranean Pollution Control', *International Organization* 43 (3): 377–403.

Haas, P. (1992) 'Introduction: Epistemic Communities and International Policy Co-ordination', *International Organization* 46 (1): 1–35.

Hall, P.A. (1983) 'Policy Innovation and the Structure of the State: The Politics-Administration Nexus in France and Britain', *Annals* 466 (March): 43–59.

Hall, P.A. (1986) *Governing the Economy*, New York: Oxford University Press.

Hall, P.A. (1993) 'Policy Paradigms, Social Learning, and the State', *Comparative Politics* 25 (3) (April): 275–96.

Harrison, M. (1993) 'The President, Cultural Projects and Broadcasting Policy', in Hayward, J.E.S., ed. *De Gaulle to Mitterrand. Presidential Power in France*, London: Hurst, 190–218.

Hartmann, H.-J. (1991) 'Zehn Jahre Kulturpolitik der französischen Sozialisten (1981–1991), *Lendemains* 16 (63): 107–16.

Hassenteufel, P. (1995) '*Do policy networks matter*? Lifting descriptif et analyse de l'Etat en interaction', in Le Galès and Thatcher, eds. (1995), 91–107.

Hauss, C. (1985) 'Culture and the state: an interview with Jack Lang', *Contemporary French Civilization* 9 (2): 213–22.

Hayward, J. (1986a) *The State and the Market Economy*, Brighton: Wheatsheaf.

Hayward, J. (1986b) 'The Nemesis of Industrial Patriotism: the French Response to the Steel Crisis', in Mény, Y. and Wright, V., eds. *The Politics of Steel: Western Europe and the Steel Industry in the crisis years (1974–1984)*, Berlin: De Gruyter, 502–33.

Hayward, S. (1993) 'State, Culture and the Cinema: Jack Lang's strategies for the French Film Industry 1981–1993', *Screen* 34 (4) (Winter): 380–91.
Heclo, H. and Wildavsky, A. (1974) *The Private Government of Public Money*, London: Macmillan.
Heclo, H. (1978) 'Issue Networks and the Executive Establishment', in King, A., ed. *The New American Political System*, Washington: American Enterprise Institute, 87–124.
Hix, S. (1994) 'The Study of the European Community: the Challenge to Comparative Politics', *West European Politics* 17 (1): 1–30.
Holleaux, A., ed. (1982) 'Les pouvoirs publics et la culture', *Revue française d'administration publique* 22 (June): 5–130.
Hunter, M. (1990) *Les jours les plus Lang*, Paris: Odile Jacob.
Inspection Générale des Finances (1982) *Rapport d'enquête sur la gestion des subventions du ministère de la Culture. Par M. Jean-Paul Cluzel, Inspecteur des Finances* (unpublished), Paris: Inspection Générale des Finances.
Jenger, J. (1986) *Orsay de la gare au musée*, Paris: Electra Moniteur.
Jobert, B. (1988) 'La version française du corporatisme', in Colas, ed. (1988): 3–18.
Jobert, B. and Muller, P. (1987) *L'Etat en action, politiques publiques et corporatismes*, Paris: PUF.
Jourdheuil, J. (1987) 'Le théâtre immobile', *Libération* (29 May 1987): 33–4.
Jordan, G. and Schubert, K. (1992) 'A Preliminary Ordering of Policy Network Labels', *European Journal of Political Research* 21 (February): 7–27.
Karmitz, M. (1994) *Bande à part*, Paris: Grasset.
Jung, C. (1987) *Le Parti socialiste et l'action culturelle: formation d'un programme de gouvernement, 1971–1981, Mémoire de DEA, Etudes Politiques* (unpublished), Paris: Institut d'Etudes Politiques.
Keeler, J.T.S. (1987) *The Politics of Neo-Corporatism in France*, New York: Oxford University Press.
Kenis, P. and Schneider, V. (1991) 'Policy Networks and Policy Analysis', in Marin and Mayntz, eds. (1991): 25–59.
Kesler, J.-F. (1985) *L'E.N.A., la société, l'Etat*, Paris: Berger-Lerrault.
Lang, J. et al. (1983) 'Comparaison entre les politiques culturelles de différents pays', *Cultures* (33): 67–80.
Lang, J. (1985a) 'Les musiciens de rock opposent un désaveu au racisme (interview)', *Ouest France* (19 March 1985).
Lang, J. (1985b) 'Quand PM fait "chanter" Jack Lang. Un entretien avec le ministre de la Culture', *Paroles et musique* 51 (June/July/August): 24–7.
Lang, J. (1987) 'Dix questions à M. Jack Lang', *Revue française d'administration publique* 44 (October/December): 45–50.
Lassalle, M. de (1993) 'La "nouvelle" politique de la lecture (1981–1986)', *Politix* 24 (4): 78–94.
Laurent, J. (1955) *La République et les Beaux-Arts*, Paris: Julliard.
Laurent, J. (1982) *Arts et pouvoirs en France de 1793 à 1981: histoire d'une démission artistique*, Saint-Etienne: CLEREC.
Lebrun, J. (1984) 'Entre socialisme et féodalisme le ministère danse', *Esprit* 87 (March): 48–50.
Le Corroller, P. (1985) 'Où va l'argent de la culture?', *Le nouvel économiste* 510 (11 October 1985): 74–80.

Le Galès, P. (1995) 'Les réseaux d'action publique entre outil passe-partout et théorie de moyenne portée', in Le Galès and Thatcher, eds. (1995), 13–28.

Le Galès, P. and Thatcher, M., eds. (1995) *Les réseaux de politique publique: débat autour des policy networks*, Paris: L'Harmattan.

Léonard, Y., ed. (1993) *Culture et société*, Paris: La Documentation française.

Léotard, F. (1988) *Culture: Les chemins de printemps*, Paris: Albin Michel.

Lion, B. (1985) *Modalités et limites de l'aide de l'Etat dans un nouveau secteur d'intervention: les 'associations-rock'*, Paris: Université Paris I.

Lion, B., Commins, D. and Karczewski, J. (1985) *Le rock à la recherche de lieux*, Paris: Ministère de la Culture, Service des Etudes et Recherches.

Lompech, A. (1990) 'Une partition pleine de fausses notes', *Le Monde* (8 March 1990): 21–2.

Looseley, D.J. (1990) 'Jack Lang and the Politics of Festival', *French Cultural Studies* 1 (1): 5–19.

Looseley, D.J. (1994) 'Cultural Policy and Democratization under Mitterrand', in Raymond, G., ed. *France During the Socialist Years*, Aldershot: Dartmouth, 117–37.

Looseley, D. (1995) *The Politics of Fun*, Oxford: Berg.

Louis, G. (1985) 'Culture: un budget doublé', *Revue politique et parlementaire* 87 (916–917): 30–9.

Lowi, T. (1964) 'American Business, Public Policy, Case Studies, and Political Theory', *World Politics* XVI (4): 677–715.

Lowi, T. (1969) *The End of Liberalism*, New York: Norton.

Lussato, B. and Messadié, G. (1986) *Bouillon de culture*, Paris: Laffont.

Marest, J. (1986) 'La culture "made in Lang"', *La pensée* 249 (February): 69–77.

Marin, B. and Mayntz, R., eds. (1991) *Policy Networks. Empirical Evidence and Theoretical Considerations*, Frankfurt/Boulder: Campus/Westview.

Marin, B. and Mayntz, R. (1991) 'Introduction: Studying Policy Networks', in Marin and Mayntz, eds. (1991): 11–23.

Marsh, D. (1995) 'Théorie de l'Etat et modèle de réseaux d'action publique', in Le Galès and Thatcher, eds. (1995): 141–63.

Marsh, D. and Rhodes, R.A.W., eds. (1992) *Policy Networks in British Government*, Oxford: Oxford University Press.

Marsh, D. and Rhodes, R.A.W. (1992) 'Policy Communities and Issue Networks: Beyond Typology', in Marsh and Rhodes, eds. (1992): 249–68.

Menger, P.-M. (1983) *Le paradoxe du musicien*, Paris: Flammarion.

Menger, P.-M. (1984) 'Création et consommation musicales: le grand écart (interview)', *Esprit* 87 (March): 51–62.

Menger, P.-M. (1985) 'L'élitisme musical républicain', *Esprit* 99 (March): 5–19.

Menger, P.-M. (1987) 'L'Etat-providence et la culture', in Chazel, ed. (1987): 41–52.

Menger, P.-M. (1989a) *Les laboratoires de la création musicale*, Paris: Ministère de la Culture, Département des études et de la prospective/La Documentation française.

Menger, P.-M. (1989b) 'L'opéra, bien public? A propos de l'Opéra-Bastille', *Esprit* 148–9 (March/April): 88–96.

Menger, P.-M. (1990) 'Transformation de la politique culturelle publique', *Commentaire* 49 (Spring): 113–16.

Menger, P.-M. (1994) 'Etre artiste par intermittence: la flexibilité du travail et le risque professionnel dans les arts du spectacle', *Travail et emploi* 60 (3): 4–22.

Menger, P.-M. and Patureau, F. (1987) *L'Art entre science et technique: le cas de la recherche musicale*, Paris: Ministère de la Culture, Département des études et de la prospective/CNRS – Centre de sociologie des arts.

Mény, Y. (1983) 'La décentralisation', *Administration* 1982, Paris: Institut international d'administration publique: 13–58.

Mény, Y. (1989) 'Formation et transformation des *policy communities*. L'exemple français', in Mény, Y., ed. *Idéologies, partis politiques et groupes sociaux*, Paris: Presses de la FNSP, 355–66.

Mesnard, A.-H. (1974) *La politique culturelle de l'Etat*, Paris: PUF.

Mesnard, A.-H. (1990) *Droit et politique de la culture*, Paris: PUF.

Miège, B. and Ion, J. (1974) *L'appareil d'action culturelle*, Paris: Editions universitaires.

Mignon, P. (1993) 'Existe-t-il une "culture rock"? (interview)', *Esprit* 193 (July): 140–50.

Mills, M. and Saward, M. (1994) 'Policy Communities: Theoretical Issues', in Dunleavy, P. and Stanyer, J., eds. *Contemporary Political Studies, 1994, vol. I*, Belfast: The Political Studies Association of the United Kingdom, 79–92.

Ministère de la Culture (1985a) *La politique culturelle 1981–1985. Bilan de la Législature,* Paris: Ministère de la Culture.

Ministère de la Culture (1985b) *Culture et recherche: rapport de conjoncture et de la prospective*, Paris: Ministère de la Culture.

Ministère de la Culture (1986a) *Atlas du changement culturel*, Paris: Editions W.

Ministère de la Culture (1986b) *Conférence internationale sue l'économie de la culture, Avignon, 1986*, 4 volumes, Paris: La Documentation française.

Ministère de la Culture, de la Communication et des Grands Travaux, (1989a) *Les grands travaux*, Paris: Ministère de la Culture, de la Communication et des Grands Travaux.

Ministère de la Culture, de la Communication et des Grands Travaux (1989b) *30e anniversaire du Ministère de la Culture: journées d'étude sur la création du Ministère de la Culture, 30 novembre–1er décembre 1989: dossier d'information*, Paris: Ministère de la Culture, de la Communication et des Grands Travaux.

Ministère de la Culture, de la Communication et des Grands Travaux (1989c) *Conférence de presse: 'Rock et variétés', lundi 25 septembre 1989* (presskit).

Ministère de la Culture, de la Communication et des Grands Travaux (1989d) *Conférence de presse de Jack Lang sur la musique, 6 novembre 1989* (presskit).

Ministère de la Culture, de la Communication et des Grands Travaux (1990) *Le Ministère de la Culture, de la Communication et des Grands Travaux*, Paris: Ministère de la Culture, de la Communication et des Grands Travaux.

Ministère de la Culture, de la Communication et des Grands Travaux (1991a) *1981–1991: Bilan de la politique culturelle.* Paris: Ministère de la Culture, de la Communication et des Grands Travaux.

Ministère de la Culture, de la Communication et des Grands Travaux (1991b) *Les musées*, Paris: La Documentation française.

Ministère de la Culture et de la Francophonie (1993) *Orientations générales de la politique culturelle: dossier de presse*, Paris: Ministère de la Culture et de la Francophonie.

Ministère de la Culture, Département des Etudes et de la Prospective (1986) *Le budget du Ministère de la Culture française – éléments pour une comparaison européenne*, Paris: Ministère de la culture et de la communication, Département des études et de la prospective.

Ministère de la Culture, Département des Etudes et de la Prospective (1987) *Les dépenses culturelles des administrations d'Etat autres que le Ministère de la culture: 1981–1984*, Paris: La Documentation française.

Ministère de la Culture, Département des Etudes et de la Prospective, ed. (1988a) *4e conférence internationale sur l'économie de la culture, vol. II: Cultures en devenir et volonté publique*, Paris: La Documentation française.

Ministère de la Culture, Département des Etudes et de la Prospective (1988b) 'Les dépenses culturelles des régions', *Développement culturel* 73 (February): 1–5.

Ministère de la Culture, Département des Etudes et de la Prospective (1989) *Les politiques culturelles des villes et leurs administrateurs*, Paris: La Documentation française.

Ministère de la Culture, Département des Etudes et de la Prospective (1991a) *Le soutien aux industries culturelles en Europe et au Canada, par François Rouet et Xavier Dupin*, Paris: La Documentation française.

Ministère de la Culture, Département des Etudes et de la Prospective (1991b) *Les dépenses culturelles des communes*, Paris: La Documentation française.

Ministère de la Culture, Département des Etudes et de la Prospective (1992a) *Les pratiques culturelles des jeunes*, Paris: La Documentation française.

Ministère de la Culture, Département des Etudes et de la Prospective (1992b) *Le marché de l'art contemporain en France*, Paris: La Documentation française.

Ministère de la Culture, Département des Etudes et de la Prospective (1992c) *Le mécénat: intervention des entreprises dans la vie culturelle en France et à l'étranger*, Paris: La Documentation française.

Ministère de la Culture, Département des Etudes et de la Prospective (1992d) *Les dépenses culturelles des départements*, Paris: La Documentation française.

Ministère de la Culture, Département des Etudes et de la Prospective (1993) *Chiffres clé 1993*, Paris: La Documentation française.

Ministère de la Culture, Mission de la recherche (1984) *La recherche au Ministère de la Culture*, Paris: Imprimerie Nationale.

Ministère de la Culture, Service des Etudes et Recherches (1986) *Politiques culturelles: études et documents, 1976–1983, par Lorraine Maicho et al.*, Paris: La Documentation française.

Ministère de l'Education nationale et de la Culture (1993) *Présentation des nouvelles mesures en faveur du cinéma: point de presse, mercredi 24 février 1993*, Paris: Ministère de l'Education nationale et de la Culture.

Mollard, C. (1982) 'Pour une culture démocratique', *Politique aujourd'hui* 1–2: 105–13.

Mollard, C. (1983) 'La décentralisation artistique. Bilan et perspectives', *Lendemains* 8 (30): 55–61.

Mollard, C. (1984) *Le mythe de Babel: l'artiste et le système*, Paris: Grasset.
Mollard, C. (1986) *La passion de l'art: écrits et paroles, 1981–1985*, Paris: La Différence.
Mollard, C. (1987) *Profession: ingénieur culturel. Manifeste pour une nouvelle manière de penser l'action culturelle,* Paris: La Différence.
Montassier, G. (1980) *Le fait culturel,* Paris: Fayard.
Mooney, M. M. (1980) *The Ministry of Culture: Connections among Art, Money and Politics*, New York: Wyndham.
Morot-Sir, E. (1983) "Vers une conscience et politique nouvelles de la culture en France", *Contemporary French Civilisation* 8 (1–2): 263–93.
Moulin, R. (1992a) *L'artiste, l'institution et le marché*, Paris: Flammarion.
Moulin, R. (1992b) 'La commande publique', *Revue des deux mondes* 11 (November): 63–70.
Moulin, R., ed. (1986) *Sociologie de l'art*, Paris: La Documentation française.
Moulinier, P. (1982) 'Désétatiser, déparisianiser la culture', *Pour* 83–84 (June): 128–34.
Moulinier, P. (1988) 'Le ministre de la Culture et ses subventions', *Revue française de finances publiques* 23: 67–77.
Moulinier, P. and Rizzardo, R. (1988) *La coopération des collectivités publiques dans le financement de l'action culturelle: des financements croisés aux financements conjoints*, Paris: Ministère de la Culture.
Mounier, G. (1991) *Des beaux-arts aux arts plastiques*, Paris: Manufacture.
Mulcahy, K. and Swaim, R., eds. (1982) *Public Policy and the Arts*, Boulder: Westview Press.
Myerscough, J., ed. (1984) *Funding the Arts in Europe*, London: Policy Studies Institute.
Nidaj, J.-M., ed. (1986) 'L'enjeu culturel', *Après-demain* 33 (February): 3–18.
Les nouvelles littèraires, ed. (1981) '7 ans de culture', *Les nouvelles littèraires* 59 (2784): 26–43.
Ory, P. (1983) *L'entre-deux-mai: histoire culturelle de la France, mai 1968–mai 1981,* Paris: Seuil.
Ory, P. (1984) 'La politique du ministère Jack Lang: un premier bilan', *The French Review* VIII (October): 77–84.
Ory, P. (1987) 'L'histoire culturelle de la France contemporaine. Question et questionnement', *Vingtième siècle* 16 (October/December): 67–82.
Ory, P. (1989) *L'aventure culturelle française: 1945–1989*, Paris: Flammarion.
Ory, P. (1996) 'L'Etat et la Culture, de la Révolution à 1959', in Perret, J. and Saez, G., eds (1996), *Institutions et vie culturelle*, Paris: La Documentation Française, 7–11.
O'Toole, L. (1993) 'Multiorganizational Policy Implementation: Some Limitations and Possibilities for Rational-Choice Contributions', in Scharpf, ed. (1993): 27–63.
Padioleau, J. (1981) *Quand la France s'enferre*, Paris: PUF.
Passeron, A. (1987) 'Comment ils ont placé leurs amis', *Pouvoirs* 40: 25–30.
Petit-Castelli, C. (1981) *La Culture à la Une ou l'Action culturelle dans les Mairies Socialistes*, Paris: Club Socialiste du Livre.
Pinto, D. (1987) 'The Left, the intellectuals and culture', in Ross, G., Hoffman, S. and Malzacher, S., eds. *The Mitterrand Experiment*, New York: Oxford University Press.

Pisier, E. (1991) 'Politique du livre et de la lecture: entretien avec Evelyne Pisier', *Esprit* 170 (March/April): 116–29.
Plunkett, P. de (1982) *La culture en veston rouge*, Paris: La Table Ronde.
Policy Studies Institute (1990) 'Funding the Arts in Seven Western Countries', *Cultural Trends* 5 (March), London: Policy Studies Institute.
Politix, ed. (1993) 'Affaires culturelles', *Politix* 24 (December).
Pommerehne, W. and Frey, B. (1993) *La culture a-t-elle un prix? Bilan de 10 ans d'enquête sur les coûts de l'art,* Paris: Commentaire/Plon.
Pongy, M. (1990) 'Municipalités en action et politiques publiques locales dans le champ de la culture', *Les cahiers du CERASS* 21 (October): 77–89.
Pontier, J.-M. (1993) 'Culture et planification', *Revue Administrative* 46 (274): 304–10.
Pontier, J.-M., Ricci, J.-C. and Bourdon, J. (1990) *Droit de la culture*, Paris: Dalloz.
Poujol, G. (1991) 'The creation of a Ministry of Culture in France', *French Cultural Studies* 2 (6): 251–60.
Poumarede, J. (1988) 'Décentralisation et pluralisme culturel', *Annales de l'Université des sciences sociales de Toulouse* 36: 153–66.
Pouvoirs locaux (1993) 'La culture malade de l'Etat', *Pouvoirs locaux* 18: 52–80.
Quermonne, J.-L. (1994) 'La "mise en examen" des cabinets ministériels', *Pouvoirs* 68 (January): 61–75.
Racine, B. (1993) 'Sur dix ans de politique culturelle', *Revue des deux mondes* 3: 32–8.
Racine, P.-F. (1994) 'Les apports des SOFICA', *ENA mensuel* (April): 18–19.
Rémy, P.-J. [pseudonym of: Angremy, J.-P.] (1989) *Bastille: rêver un opéra*, Paris: Plon.
Renard, J. (1987) *L'élan culturel: La France en mouvement*, Paris: PUF.
Revue française d'administration publique, ed. (1993) 'Administrer la culture?' *Revue française d'administration publique* 65 (January/March): 5–130.
Revue française d'administration publique, ed. (1987) 'L'audiovisuel: enjeux et paradoxes', special issue of *Revue française d'administration publique* 44 (October/December): 5–144.
Rhodes, R.A.W. (1981) *Control and Power in Central-Local Government Relations*, Farnborough: SSRC/Gower.
Rhodes, R.A.W. (1985) *Power-Dependence, Policy Communities and Intergovernmental Relations*, Essex papers in Politics and Government no. 30, Colchester: University of Essex.
Rhodes, R.A.W. (1986) *The National World of Local Government*, London: Unwin Hyman.
Rhodes, R.A.W. and Marsh, D. (1992) 'New Directions in the Study of Policy Networks', *European Journal of Political Research* 21: 181–205.
Richardson, J.J. and Jordan, G. (1979) *Governing under Pressure*, Oxford: Basil Blackwell.
Rigaud, J. (1975) *La Culture pour Vivre*, Paris: Gallimard.
Rigaud, J. (1986) 'Pouvoir et non-pouvoir du ministre', *Pouvoirs* 36: 5–14.
Rigaud, J. (1989) 'La culture et le pouvoir', *Revue des deux mondes* 4: 61–70.
Rigaud, J. (1990) *Libre Culture,* Paris: Gallimard.
Rigaud, J. (1995) *L'Exception culturelle*, Paris: Grasset.

Rigaud, J. and Delcros, X. (1986) *Les institutions administratives françaises. Vol. II: Le fonctionnement*, Paris: Presses de la FNSP/Dalloz.

Rigby, B. (1991) *Popular Culture in Modern France*, London: Routledge.

Rioux, J.-P. (1987) *L'histoire culturelle de la France contemporaine. Bilans et perspectives de la recherche* (unpublished), Paris: Ministère de la Culture et de la Communication.

Rioux, J.-P. (1992) "L'Etat culturel depuis la Libération", *Le Débat* 70 (May/August): 60–5.

Ritaine, E. (1983) *Les Stratèges de la Culture*, Paris: Presses de la FNSP.

Rizzardo, R., ed. (1993) 'Le rôle des collectivités locales – Débat animé par M. René Rizzardo . . . ', *Revue française d'administration publique* 65 (January/March): 73–87.

Rizzardo, R. and Moulinier, P. (1990) *La décentralisation culturelle. Rapport au ministre de la Culture*, Paris: La Documentation française.

Roman, J. (1991) 'Controverse: contre l'Etat culturel de Marc Fumaroli?' *Esprit* 176 (October): 149–57.

Rouet, F. (1987) *Des aides à la culture*, Brussels: Edition Mardaga.

Roux, B. (1993) *L'économie contemporaine du spectacle vivant*, Paris: Harmattan.

Roux, E. de (1993) 'La course de Jack Lang', *Le Monde* (22 February 1993).

Ruby, C., Nouvel, K. and Simonet, J. (1993) 'La bataille du culturel', *Regards sur l'actualité* 189: 3–19.

Saez, G. (1985a) 'Les politiques de la culture', in Grawitz, M. and Leca, J., eds. *Traité de science politique, vol. IV*, Paris: PUF, 387–422.

Saez, G. (1985b) 'Politique culturelle: suivez le guide!', *Pour* 101 (June): 36–45.

Saez, G. (1988) 'Modernisation et corporatisme dans la politique culturelle française', in Colas, ed. (1988): 91–107.

Saez, G., ed. (1990) *Recherche, évaluation dans les politiques culturelles. Actes du séminaire de Grenoble, 13–14 juin 1989*, Grenoble: CERAT.

Saez, G. (1993) 'Vers la fin de l'Etat culturel?', *Revue française d'administration publique* 65: 51–62.

Safran, W. (1985) 'The Mitterrand regime and its policies of ethnocultural accommodation', *Comparative Politics* 18 (October): 41–63.

Sagot-Duvauroux, D. (1985) *Structure de financement et organisation d'un système: l'exemple du théâtre, thèse troisième cycle* (unpublished), Paris: Université Paris I.

Sagot-Duvauroux, D. (1988) 'Le marché de la subvention publique au théâtre: du monopsone au monopole', in Ministère de la Culture, Département des Etudes et de la Prospective, ed. (1988a): 229–43.

Sagot-Duvauroux, D. (1992) 'Les procédures de négociation des subventions aux organisations culturelles: application de deux modèles économiques', *Revue de l'économie sociale* 29: 63–72.

Saint-Pulgent, M. de (1991) *Le syndrome de l'Opéra*, Paris: Laffont.

Scharpf, F.W. (1993) 'Coordination in Hierarchies and Networks', in Scharpf, ed. (1993): 125–65.

Scharpf, F.W., ed. (1993) *Games in Hierarchies and Networks*, Frankfurt/Boulder, CO: Campus/Westview.

Schneider, M. (1990) 'Entretien avec Michel Schneider', *Le Monde de la Musique* 130 (February): 56–7.

Schneider, M. (1993a) *La Comédie de la Culture*, Paris: Seuil.

Schneider, M. (1993b) 'D'une culture privée de public', *Pouvoirs locaux* 18 (October): 54–60.

Schneider, V. and Werle, R. (1991) 'Policy Networks in the German Telecommunications Domain', in Marin and Mayntz, eds. (1991): 97–136.

Schuster, J.M.D. (1984) *Influences des mesures d'incitation fiscales sur le secteur artistique en Europe occidentale*, New Haven: ISPS, Yale University.

Schuster, J.M.D. (1985a) *International Comparative Study of Arts Support: Canada, Federal Republic of Germany, France, Italy, Great Britain, Netherlands, Sweden, United States*, Washington: National Endowment for the Arts.

Schuster, J.M.D. (1985b) *Méthodes pour l'évaluation des politiques culturelles nationales: séminaire, 16–18 avril 1985* (unpublished), Paris: Ministère de la Culture, Département des Etudes et de la Prospective.

Schuster, J.M.D. (1985c) *Program Evaluation and Cultural Policy: Working Paper*, Cambridge: Massachusetts Institute of Technology.

Sinz, D. (1983) 'Jack Lang – Kultur als Staatsaffäre. Sein Versuch einer "Kultur-Revolution à la française"', *Dokumente* 39 (1): 38–43.

Stimpson, B. and Somorjay, S., eds. (1991) *La politique culturelle sous Mitterrand*, London: Department of Modern Languages, Roehampton Institute.

Street, J. (1994) *Passionate Interests: The Politics of Popular Culture*, London: Polity Press.

Suleiman, E. (1974) *Politics, Power and Bureaucracy in France*, Princeton: Princeton University Press.

Suleiman, E. (1987) *Private Power and Centralization in France: The Notaires and the State*, Princeton: Princeton University Press.

Tasca, C. (1981) 'Démocratisation culturelle et maisons de la culture', in Reynaud, J.-Y. and Grafmeyer, Y., eds. *Français qui êtes-vous?*, Paris: La Documentation française, 435–44.

Teillet, P. (1992) *Le discours culturel et le rock. L'expérience des limites de la politique culturelle de l'Etat*, *Thèse de doctorat* (unpublished), Rennes: Université Rennes I.

Thatcher, M. (1995) 'Les réseaux de politique publique: bilan d'un sceptique', in Le Galès and Thatcher, eds. (1995), 229–48.

Turc, S. (1987) *L'Etat et la culture: le ministère des Affaires culturelles de 1959 à 1969*, Grenoble: Universitè des sciences sociales de Grenoble.

UNESCO (1981) *La politique culturelle en France: l'action du Ministère de la Culture et de la Communication*, Paris: UNESCO.

Urfalino, P. (1984) *L'allocation de ressources sans critères de choix: la mise en oeuvre des politiques culturelles municipales, Thèse trosième cycle: sociologie* (unpublished), Paris: Institut d'Etudes Politiques.

Urfalino, P. (1987) 'Media, performers and politicians: strategies of delocalization', *Media, Culture and Society* 9 (3): 333–45.

Urfalino, P. (1989) 'Les politiques culturelles: mécénat caché et académies invisibles', *L'année sociologique* 39 (1989): 81–110.

Urfalino, P. (1990) *Quatre voix pour un opéra*, Paris: Métaillié.

Urfalino, P. (1993a) 'Action publique, création privée', *Revue française d'administration publique* 65: 51–62.

Urfalino, P. (1993b) 'De l'anti-impérialisme américain à la dissolution de la politique culturelle', *Revue française de science politique* 5 (October): 823–49.

Urfalino, P. (1993c) 'La philosophie de l'Etat esthétique', *Politix* 24 (4): 20–35.
Urfalino, P. (1997) 'Quelles missions pour le ministère de la Culture?', *Esprit* 228 (January): 37–59.
Van Waarden, F. (1992) 'Dimensions and Types of Policy Networks', *European Journal of Political Research* 21: 29–52.
Verheuge, R. (1988) *Gérer l'utopie: l'action culturelle dans la cité*, Aix-en-Provence: Edisud.
Villiers, P. de (1986) 'La fin de la culture d'Etat', *Revue des Deux Mondes* 10: 22–9.
Von Beyme, M. (1990) *Kulturpolitik unter den Sozialisten in Frankreich von 1981 bis 1986*, Frankfurt: Lang.
Wachtel, D. (1987) *Cultural Policy and Socialist France*, Westport: Greenwood Press.
Wais, A. (1984) 'Rock around Jack Lang', *Le Monde* (19 December 1984).
Wallon, E., ed. (1991) *L'artiste, le prince. Pouvoirs publics et création*, Grenoble: Musée de la civilisation de Québec/Presses universitaires de Grenoble.
Wangermée, R. (1988) 'Rapport du groupe d'experts européens', in Conseil de l'Europe, Conseil de la coopération culturelle, Programme Européen d'Evaluation, ed. (1988): 9–273.
Wellman, B. and Berkowitz, S., eds. (1988) *Social Structures: a Network Approach*, Cambridge: Cambridge University Press.
Wilks, S. and Wright, M. (1987) 'Conclusion: Comparing Government-Industry Relations: States, Sectors and Networks', in S. Wilks and M. Wright, eds. *Comparative Government-Industry Relations*, Oxford: Clarendon Press, 274–311.
Wilson, F. (1987) *Interest Group Politics in France*, Cambridge: Cambridge University Press.
Worms, J.-P. (1966) 'Le préfet et ses notables', *Sociologie du travail* 8: 249–75.
Wright, V. (1989) *The Government and Politics of France*, London: Unwin Hyman.
Yonnet, P. (1985) *Jeux, modes et messes: la société française et le moderne: 1945–1985*, Paris: Gallimard.
Zintl, R. (1993) 'Commentary on O'Toole and Scharpf: Networks as a Challenge for Theory', in Scharpf, ed. (1993): 177–86.

Index